Bee Pollination in Agricultural Ecosystems

Bee Pollination in Agricultural Ecosystems

Edited by
Rosalind R. James and Theresa L. Pitts-Singer

UNIVERSITY PRESS

2008

OXFORD
UNIVERSITY PRESS

Oxford University Press, Inc., publishes works that
further Oxford University's objective of excellence
in research, scholarship, and education.

Oxford New York
Auckland Cape Town Dar es Salaam Hong Kong Karachi
Kuala Lumpur Madrid Melbourne Mexico City Nairobi
New Delhi Shanghai Taipei Toronto

With offices in
Argentina Austria Brazil Chile Czech Republic France Greece
Guatemala Hungary Italy Japan Poland Portugal Singapore
South Korea Switzerland Thailand Turkey Ukraine Vietnam

Copyright © 2008 by Oxford University Press, Inc.

Chapters 1, 4, 6, 7, 8, and 13 were prepared by the authors as part of their
official duties as U.S. government employees, and are in the public domain.

Published by Oxford University Press, Inc.
198 Madison Avenue, New York, New York 10016
www.oup.com

Oxford is a registered trademark of Oxford University Press

All rights reserved. No part of this publication may be reproduced,
stored in a retrieval system, or transmitted, in any form or by any means,
electronic, mechanical, photocopying, recording, or otherwise,
without the prior permission of Oxford University Press.

Library of Congress Cataloging-in-Publication Data

Bee pollination in agricultural ecosystems / edited by Rosalind R. James and
Theresa L. Pitts-Singer.
    p. cm.
Includes bibliographical references and index.
ISBN 978-0-19-531695-7
1. Bees—Control—Environmental aspects—United States.   2. Pollination
by insects.   3. Agricultural ecology—United States.   I. James, Rosalind R.
II. Pitts-Singer, Theresa L.
QK926.B34 2008
577.5'5—dc22   2007043146

9 8 7 6 5 4 3 2 1

Printed in the United States of America
on acid-free paper

FOREWORD

It was not until the eighteenth century that the subject of this book, the pollination services of bees, began to be understood and valued. Nevertheless, the association between man and bees has been long and close, and dates from at least 2400 BC. Beekeeping with the Western honey bee, *Apis mellifera*, was a well-developed craft in ancient Egypt during the fifth dynasty of the Old Kingdom. When Christopher Columbus and his companions landed in Cuba in 1492, the local inhabitants greeted them with gifts of honey from a local native stingless honey bee, *Melipona beecheii*, which was, and still is, managed in log hives by native peoples in the neotropics.

Man's close association with bees led to a remarkable cultural convergence between two of the great dynastic cultures: in ancient Egypt, the hieroglyph of a honey bee was a symbol of royalty, and for the Mayans of Central America a pictograph of a stingless bee was a symbol of kingship.

It is easy to see why this should be. On both sides of the Atlantic Ocean, honey and hive products such as wax and propolis from social bees were and are important commodities in human commerce, both as food and a source of cosmetic and medicinal substances. Together with fermented honey drinks, these honey bee and stingless bee products had immediate and obvious value and made it inevitable that apiculture would evolve into a respected craft that would often be accorded religious and magical status.

We can surmise, however, that an unwitting relationship with bees dates back even further. When our immediate ancestors embarked on the evolutionary path to bipedalism and a hunter-gatherer lifestyle, they could do so because of a savannah biotope with structural and floral features resulting from the coevolved interactions between those keystone mutualists *extraordinaire*, bees and flowering plants.

Now we are mostly *not* hunter-gatherers and we have produced new habitats, the agroecosystems, where intensive agriculture has resulted in crop yields undreamed of a couple of generations ago. In so doing, we have not emancipated ourselves from dependency on bees: we rely on them to pollinate 63 of the 82 (77%) most valuable crops. Worldwide, bees pollinate more than 400 crop species and in the United States more than 130 crop species.

This is not without some irony. While we still depend mightily on the pollination services of bees, we have devastated natural floras and insect faunas in creating vast, structurally uniform monocultures that are far from bee-friendly. Hence, the next stage in the evolution of man's close relationship with bees: the migratory beekeeper. In the ecological disaster zone known as California's Central Valley, vast numbers of honey bee colonies from as far away as Texas and Florida are trucked in each year for the pollination of almonds. In 1994 this involved the rental of 1.4 million honey bee colonies. By 2012, it is estimated that 2 million colonies will be needed for the ever-expanding acreage devoted to almonds alone. At the time of this writing, there are about 2.9 million honey bee colonies in North America, of which 2–2.5 million are rented out for pollinating 13 crops.

The arithmetic of future needs, therefore, doesn't quite add up. This, together with the fact that honey bees are under growing pressure from parasites, disease, and colony collapse disorder, has understandably led to the search for additional native bee species as alternatives to honey bees as managed pollinators.

It is to the wildlands and their floras that we must look for new pollinator species, and this is happening. However, we must conserve these reservoirs, many of which are under pressure from urban sprawl and agriculture. To do this, we need to develop a greater understanding at the community level of the dynamic network of relationships between bees and flowering plants. This is not simply for the economic benefits of potential pollinators: we also accord aesthetic and recreational value to our wildlands.

Research on the nesting biology and management of native, solitary bees for specific crops is now a growing field. Moreover, we can enhance our use of the pollination services of these bees by attempting to overcome corporate agriculture's horror at the prospect of stands of native flora as supplementary forage in the vicinity of their crops.

Biosecurity can be regarded as a recurrent theme in this book, whether it is concern about pollen transfer from genetically modified crops to related weed species, mediated by the foraging movements of bees, or the unforeseen and detrimental interactions between invasive plants and native bee faunas. Unforeseen and adverse ecological effects also occur when bee species and/or subspecies are moved outside of their natural ranges. The best known example of this is the problem of "Africanized" honey bees, when bees from sub-Saharan Africa were introduced into Brazil, crossed with European honey bees (also nonnative), resulting in a multiplicity of well-known problems. Problems also have occurred with the commercial management of bumble bees for greenhouse crops, where subspecies of *Bombus terrestris* have been introduced outside of their natural range, and now cases in England and Israel document the escape and establishment of populations into the wild, with adverse effects on local bee faunas.

The above issues are outlined and discussed in this book. These are pressing matters, and this volume is therefore timely, not least because its contributors are leading current thinkers and researchers in the field. Collectively, the subjects they address indicate the broad front of future research that is necessary if we are to consolidate our relationship with bees and their sustainable exploitation and management.

The agenda is therefore set, and we will succeed. We have to. Otherwise, under "any other business," ecologically-minded people of my generation might well ask, "How will my grandchildren cope with the food riots?"

Christopher O'Toole
Sileby, Leicestershire
England
Honorary Research Associate,
Hope Entomological Collections
Oxford University Museum of Natural History;
Science Director Almond Pollination Company

ACKNOWLEDGMENTS

We extend our gratitude to the publisher, fellow scientists, and industry customers for their encouragement in the production of this book. We deeply appreciate the diligence, cooperation, and patience of all of the authors who did a fabulous and timely job of writing and revising their chapters. Moreover, we thank the chapter reviewers for their sincere and helpful comments, criticisms, suggestions, and edits. Our efforts were expedited by assistance from Agriculture Research Service technicians Ellen Klinger and Ellen Klomps and administrative assistant Amber Whittaker in ensuring the uniformity and organization of the book. We also appreciate the diligence of our research assistants, who made it possible for our research projects to continue, even as we executed the production of this book. Finally, we thank our spouses for their loving support.

CONTENTS

Forword   v
Christopher O'Toole

Contributors   xiii

Part 1: Bee-Provided Delivery Services

Chapter 1 Bees in Nature and on the Farm   3
Theresa L. Pitts-Singer and Rosalind R. James

Chapter 2 Crop Pollination Services From Wild Bees   10
Claire Kremen

Chapter 3 Crop Pollination in Greenhouses   27
José M. Guerra-Sanz

Chapter 4 Pollinating Bees Crucial to Farming Wildflower Seed for U.S. Habitat Restoration   48
James H. Cane

Chapter 5 Honey Bees, Bumble Bees, and Biocontrol: *New Alliances Between Old Friends*   65
Peter G. Kevan, Jean-Pierre Kapongo, Mohammad Al-mazra'awi, and Les Shipp

Part 2: Managing Solitary Bees

Chapter 6   Life Cycle Ecophysiology of *Osmia* Mason Bees Used as Crop Pollinators   83
Jordi Bosch, Fabio Sgolastra, and William P. Kemp

Chapter 7   Past and Present Management of Alfalfa Bees   105
Theresa L. Pitts-Singer

Chapter 8   The Problem of Disease When Domesticating Bees   124
Rosalind R. James

Part 3: Environmental Risks Associated With Bees

Chapter 9   Environmental Impact of Exotic Bees Introduced for Crop Pollination   145
Carlos H. Vergara

Chapter 10   Invasive Exotic Plant-Bee Interactions   166
Karen Goodell

Chapter 11   Estimating the Potential for Bee-Mediated Gene Flow in Genetically Modified Crops   184
James E. Cresswell

Chapter 12   Genetically Modified Crops: Effects on Bees and Pollination   203
Lora A. Morandin

Chapter 13   The Future of Agricultural Pollination   219
Rosalind R. James and Theresa L. Pitts-Singer

Index   223

CONTRIBUTORS

Mohammad Al-mazra'awi, Department of Environmental Biology, University of Guelph, Ontario, Canada

Jordi Bosch, Ecologia–Centre de Recerca Ecològica i Aplicacions Forestals (CREAF), Universitat Autònoma de Barcelona, Bellaterra, Spain

James H. Cane, U.S. Department of Agriculture–Agricultural Research Service (USDA-ARS) Pollinating Insects Biology, Management, and Systematics Research Unit, Logan, Utah

James E. Cresswell, School of Biosciences, University of Exeter, UK

Karen Goodell, Evolution, Ecology, and Organismal Biology, Ohio State University, Newark, Ohio

José M. Guerra-Sanz, Centro de Investigación y Formación Agrícola (CIFA) La Mojonera, Instituto Andaluz de Investigación y Formación Agraria, Pesquera, Alimentaria y de la Producción Ecológica (IFAPA), La Mojonera, Almeria, Spain

Rosalind R. James, U.S. Department of Agriculture–Agricultural Research Service (USDA-ARS) Pollinating Insects Biology, Management, and Systematics Research Unit, Logan, Utah

Jean-Pierre Kapongo, Greenhouse and Processing Crops Research Centre, Agriculture and Agri-Food Canada, Harrow, Ontario, Canada

William P. Kemp, U.S. Department of Agriculture–Agricultural Research Service (USDA-ARS) Red River Valley Agricultural Research Center, Fargo, North Dakota

Peter G. Kevan, Department of Environmental Biology, University of Guelph, Ontario, Canada

Claire Kremen, Department of Environmental Science Policy and Management, University of California, Berkeley

Lora A. Morandin, Department of Environmental Policy and Management, University of California, Berkeley

Theresa L. Pitts-Singer, U.S. Department of Agriculture–Agricultural Research Service (USDA-ARS) Pollinating Insects Biology, Management, and Systematics Research Unit, Logan, Utah

Fabio Sgolastra, Dipartimento di Scienze e Tecnologie Agroambientali, Area Entomologia, Università di Bologna, Italy

Les Shipp, Greenhouse and Processing Crops Research Centre, Agriculture and Agri-Food Canada, Harrow, Ontario, Canada

Carlos H. Vergara, Departamento de Ciencias Químico-Biológicas, Universidad de las Américas-Puebla, Cholula, Puebla, Mexico

Part 1

Bee-Provided Delivery Services

1 Bees in Nature and on the Farm

Theresa L. Pitts-Singer and Rosalind R. James

Introduction

When we say that we work at "the Bee Lab," most people automatically imagine us decked out in bee suits, standing next to box-shaped hives, and holding our breaths amidst a barrage of honey bees. Although our research facility is one of five U.S. Department of Agriculture's Agricultural Research Service laboratories that are dedicated to bee research, our focus is unique in its emphasis on non-honey bees that are important or potential pollinators. The other four U.S. bee research facilities focus on honey bees, examining various aspects of honey bee biology, pest control, management, and pollination. As we have grown in our understanding of the importance of a variety of bees as pollinators in agricultural systems, we have been inspired to compile this book.

This book illustrates the importance of both managed and wild bees in agricultural ecosystems. For much of agriculture, the vital role that pollinators play in successful crop or seed production is clear and direct. Commercially managed bees are available for pollination services and are used in large commercial fields, small gardens, or enclosures such as greenhouses and screen houses. Although the general public gives honey bees much of the pollination credit, managed bumble bees and solitary bees also have made a great impact on certain commodities, and wild bees provide free pollination services that often go unnoticed. However, all of these bees are valuable and significant in their liberal passing of pollen from one plant to the next. With the recent concern over the unexplained loss of honey bee colonies, referred to as colony collapse disorder, it seems ever more important to highlight some of the other bees that could be managed for crop pollination.

Bees have also evolved a variety of social systems. The bees commonly used for pollination fall into the categories of highly eusocial, primitively eusocial, and solitary. Eusocial insects include all ants, some wasps and bees, and termites. Eusocial hymenoptera are defined by three criteria: (1) only one or a few females in a colony reproduce; (2) the colony consists of individuals from overlapping generations, including one or more queens plus her daughters and sons; and (3) brood care is cooperative within the colony. No single theory alone suffices to explain how eusociality evolved or how it is maintained. It has been theorized that altruistic, cooperative behavior can be explained by a close relatedness among cooperating individuals, but such a theory does not fully explain the social complexity of insects that live in colonies, because not all social insect societies are composed of close relatives. Reproduction and cooperation in the colony is usually controlled by the queen. For the highly eusocial honey bees, the queen maintains reproductive dominance over the workers by producing a chemical compound called queen pheromone. In the primitively eusocial bumble bees, queen pheromonal control is not well developed, and aggressive behavior toward other egg layers is the prevailing enforcement strategy (Michener, 1974).

In honey bees, the female queen and workers are strikingly different in behavior, physiology, and morphology. The queen honey bee would die if left without workers because she is designed only for mating and reproducing, not for foraging and brood care. Honey bee colonies also are long-lived and store food for colony members to use during times of dearth or inclement weather and during the winter. Primitively eusocial bumble bee queens are structurally equivalent to workers but are larger in size. Unlike honey bees, bumble bee queens live alone at the beginning of the colony cycle, foraging and taking care of the brood until the first workers emerge. Bumble bees store small amounts of honey and pollen for adults and brood, but the colony is usually short-lived and does not persist through the winter (Michener, 1974; Heinrich, 1979). Only the new generation of reproductive females, already mated, hibernates over the winter period. However, colonies of the European bumble bee *Bombus terrestris* have persisted throughout the recently milder winter months in England and New Zealand, which may demonstrate phenological plasticity in this bee species (Goulson, 2003). Other primitively eusocial bees include many sweat bees (Halictinae) and carpenter bees (Xylocopinae; Michener, 1974).

In many respects, each female solitary bee is both a "queen" and a "worker" at the same time. Solitary bees do not form colonies and have no social colony structure. A solitary female constructs her own nest and then provides food for each of her brood in the form of a mass of pollen and nectar. After this, she usually dies or departs without further care to her young and before her offspring complete their development. As a result, there is no chance for cooperation between mothers and daughters. The adult life of these bees is short, spanning only a matter of weeks. Solitary bees may nest alone, or they may nest in aggregations. Commonly, nest aggregations occur among bees that nest in the soil, but some cavity-nesting bees will form aggregations if nesting sites are available, as is common with alkali bees, mason bees, and leafcutting bees (Michener, 1974). The tendency for some solitary bees to form nest aggregations makes them particularly amenable to management for agriculture because it allows farmers to provide concentrated nest sites for the bees.

An Industrious Hum

Why are bees such notoriously hard-working pollinators? Bees are the ultimate pollinators. They are superior to other flower-visiting insects in pollination efficacy for many crops because of their abundant pollen-trapping body hair, specialized flower-handling and foraging behaviors, and reliance on floral rewards for raising offspring (Free, 1993). What benefits are gained from plants by the bees? For bees, it is all about collecting pollen and nectar and, in some cases, essential oils. These rewards are produced by plants and collected by bees as food for their brood and as fuel for adult activities. What benefits are gained from bees by the plants? For flowering plants, it is all about improving reproduction and spreading their genes. The plants benefit when bees come into contact with the reproductive structures of flowers. Bee activity increases pollen movement because the bees transport pollen grains from flower to flower and from plant to plant, delivering them to receptive stigmas and providing for cross-pollination.

Most people think of pollen as a larval bee food that is high in protein. Indeed, pollen contains 16–60% protein, but also can be a source of fats, starches, sugar, phosphates, vitamins, and sterols (Standifer et al., 1968; Svoboda et al., 1983; Buchmann, 1986; Barth, 1991; Proctor et al., 1996). Most flowers offer both pollen and nectar, but some offer only pollen. Millions of pollen grains are available per flower for a bee to collect, and such an abundance of pollen grains creates an ample resource of genetic material for plant propagation, in addition to being a source food for the propagation of plant pollinators (Barth, 1991; Proctor et al., 1996).

Nectar is commonly thought of as a carbohydrate reward offered to pollinators. Sugar (15–75%) and water are the main nectar ingredients, but other nutrients are also present, including amino acids, proteins, organic acids, phosphates, vitamins, and enzymes. Unlike pollen, nectar is not transferred between flowers by bees and plays no direct role in a plant's reproduction. However, because nectar attracts insects to flowers and is a vital component of larval provisions, nectar indirectly promotes pollination. Some plants produce oils as floral rewards that are collected by bees. Oil-collecting species include solitary bees in the families Andrenidae, Anthophoridae, and Mellitidae, and also the orchid bees (Apidae: Apinae: Euglossini). Depending on the species, these bees collect oils to mix with pollen (and may also add nectar) as a rich, fatty food source for larvae. Some bees may use oils in making water-resistant cell linings. Male euglossine bees collect highly scented flower oils from various orchids and use them to facilitate their own attractiveness to mates (Proctor et al., 1996; Roubik & Halson, 2004), and while collecting oils, these males serve as pollinators.

Consideration of Bees' Needs

For maximized production of many crops, bee pollination is required as part of a complete management system. The most successful pollination will occur when crop managers implement strategies that consider the needs of the bees. In most agricultural systems that require insect pollination, bees cannot be treated like a field application of fertilizer or herbicide. Whether managed or naturally occurring, bees need food and safe harbor for living and reproducing. If the needs of bees are met, then thriving pollinator populations will be available to provide their services year after year.

For managed bees, the timing of bee release onto a crop must co-occur with bloom so that resources are available for the bees and so that timely pollination occurs. If the crop is not in bloom when bees are ready for release, other flowering plants could be provided to maintain bees until the onset of crop bloom, or other strategies need to be taken. For example, managed solitary bees can be chilled for a short period to delay their development and facilitate timing their emergence with bloom. Failure to provide a floral resource for active bees decreases the reproductive success of bees and, in some cases, may cause them to leave the crop vicinity in search of alternative forage. Additionally, providing adequate and desirable nesting places will promote better bee retention and reproduction. Inundating a crop with pollinators may guarantee maximum crop pollination, but using an alternative method—a lower, sustainable number of foraging bees—may reduce competition for food and nesting resources, bringing about both high crop yield and greater bee reproduction rates.

Managing bees can be problematic due to the dynamics of rearing organisms in close proximity and in controlled situations. Disease epidemics can devastate or impair the production of commercial pollinators, and research is ongoing on bees that have been used domestically for thousands of years (such as the honey bee), on bees managed for decades (such as alfalfa leafcutting bees and bumble bees), and on bees on the brink of commercial-scale use (such as the red mason bee and the blue orchard bee).

In addition to managed bees, it is important to recognize the potential benefits of wild, native bees in crop systems. Mark Winston offers an encompassing view of how managerial practices might discourage wild bee pollination: "The reason that wild bees no longer visit crops are few and clear: pesticides, lack of floral diversity, habitat destruction, and, ironically, competition with managed pollinators" (Winston, 1997, 119–120).

Although wild bees are not managed, especially not in the same way as commercial bees, certain practices on or near the farm can encourage their availability and likelihood of flower visitation. If native bees already are performing a pollination service in a crop system, the addition of managed pollinators may cause competition for food and nest sites, which could result in reduction or elimination of natural pollination. Wild bumble bees, carpenter bees, sweat bees, mason bees, and other bees will visit crop flowers if favorable habitats occur in the vicinity. Favorable habitats are those in which food and safe nest sites can be found. Whether naturally or artificially created, bare patches of undisturbed ground or persistent embankments may increase aggregations of ground-nesting bees, such as alkali bees and sweat bees. Old wooden structures, loose debris piles, and thick underbrush may be attractive to carpenter bees and bumble bees as nest sites. Old, pithy plant stems, hollow reeds, or boards with drilled holes may be inviting to cavity-nesting leafcutting and mason bees. Naturally occurring flowers or deliberately added flowering plants may provide alternative sources for pollen and nectar to keep bees in a production area before the onset of crop bloom or after it has passed.

Thus, when the more preferred crop flowers are available, the bees will be available to pollinate; when the crop flowers disappear, the bees can complete their nesting.

Conclusion

Bees are extremely vital to the well-being of mankind. Products from pollinated plants, including fruits, vegetables, and seed crops, not only feed people but also feed the pets and livestock that people raise for pleasure and consumption. An appreciation of the vital relationships between plants and their pollinators, in their own time and space, is needed to secure the future of crop production. The chapters in this book are intended to provide valuable information and forethought for understanding the impact of bees in the dynamic agricultural ecosystem of modern society.

References

Barth, F. G. (1991). *Insects and flowers: The biology of a partnership*. Princeton, NJ: Princeton University Press.

Buchmann, S. L. (1986). Buzz pollination in angiosperms. In C. E. Jones & R. J. Little (Eds.), *Handbook of experimental pollination biology* (73–113). New York: Van Nostrand Reinhold.

Free, J. B. (1993). *Insect pollination of crops* (2nd ed.). London: Academic Press.

Goulson, D. (2003). *Bumblebees: Behaviour and ecology*. Oxford: Oxford University Press.

Grimaldi, D., & Engel, M. S. (2005). *Evolution of the insects*. Cambridge, UK: Cambridge University Press.

Heinrich, B. (1979). *Bumblebee economics*. Cambridge, MA: Harvard University Press.

Krombein, K. V., Hurd, P. D., Smith, D. R., & Burks, B. D. (1979). *Catalog of Hymenoptera in America north of Mexico* (Vol. 2). Washington, DC: Smithsonian Institution Press.

Michener, C. D. (1974). *The social behavior of the bees*. Cambridge, MA: Belknap Press.

———. (2000). *The bees of the world*. Baltimore: Johns Hopkins University Press.

Michener, C. D., McGinley, R. J., & Danforth, B. N. (1994). *The bee genera of North and Central America*. Washington, DC: Smithsonian Institution Press.

Proctor, M., Yeo, P., & Lack, A. (1996). *The natural history of pollination*. Portland, OR: Timber Press.

Roubik, D. W., & Halson, P. E. (2004). *Orchid bees of tropical America: Biology and field guide*. Santo Domingo de Heredia, Costa Rica: Instituto Nacional de Biodiversidad.

Standifer, L. N., Devys, M., and Barbier, M. (1968). Pollen sterols–A mass spectrographic survey. *Phytochemistry 7*, 1361–1365.

Svoboda, J. A., Herbert Jr., E. W., Lusby, W. R., and Thompson, M. J. (1983). Comparison of sterols of pollens, honeybee workers, and prepupae from field studies. *Archives of Insect Biochemistry and Physiology 1*, 25–31.

Tepedino, V. J. (2003). What's in a name? The confusing case of the Death Camas bee, *Andrena astragali* Viereck & Cockerell (Hymenoptera: Andrenidae). *Journal of the Kansas Entomological Society, 76*(2), 194–197.

Winston, M. L. (1997). *Nature wars: People vs. pests*. Cambridge, MA: Harvard University Press.

2 Crop Pollination Services From Wild Bees

Claire Kremen

Introduction

Historically, crop pollination needs were met by wild pollinators living within the farming landscape (Kevan & Phillips, 2001), and this is still true in less intensive agricultural systems (e.g., Ricketts et al., 2004; Morandin & Winston, 2005). For many modern crops requiring an animal pollinator, however, pollination is now managed as intensively as other aspects of agriculture by bringing large numbers of commercial pollinators directly to the field where pollination is needed.

Only a dozen species have been commercialized for use as pollinators (Parker et al., 1987; Batra, 2001), although thousands more species, primarily bees, participate in crop pollination (Nabhan & Buchmann, 1997). The most widely used pollinator, and the one with the longest history of domestication, is the honey bee, *Apis mellifera* (Crane, 1990), probably utilized for at least 90% of managed pollination services (Calderone, personal communication, 2005). The extent of our reliance on this single species for such an important service is risky. In the United States, managed stocks of the honey bee have declined by 50% over the past 50 years (National Research Council, 2007) due primarily to the mite, *Varroa destructor* (Morse & Goncalves, 1979; Beetsma, 1994), which both weakens individuals and transmits disease. Also, *Varroa* mites have developed resistance to the miticides (Elzen & Hardee, 2003), leading to high rates of over-winter colony mortality during some years (e.g., up to 50% across large areas of the United States), and thus high within- and between-year variability in the honey bee supply (National Research Council, 2007). *Varroa* has affected honey bee availability not only in the United States but also in Europe and the Middle East (Griffiths, 1986; Komeili, 1988).

There are two nonexclusive alternatives to our overreliance on the honey bee: domestication and commercialization of additional species (Parker et al., 1987; Kevan et al., 1990), and conservation and enhancement of populations of wild pollinators on or near farms (Batra, 2001). This chapter is concerned with the latter alternative.

Services Provided by Wild Bee Communities

We do not know how many unmanaged species contribute to crop pollination, nor what percentage of crop pollination results from visits by unmanaged species. Bees are the most important pollinators of many crops and are recorded visitors to 73% of the crop species that require pollinators worldwide (Nabhan & Buchmann, 1997). Thousands of bee species visit crop plants globally (Free, 1993), but few exhaustive surveys have been conducted. In northeastern North America alone, 190 species of bees are associated with lowbush blueberry (Kevan et al. 1990). In a single location in California, workers recorded 66 bee species visiting selected spring and summer crops (Kremen et al., 2002a). Other wild visitors to crops include flies, wasps, butterflies, moths, midges, thrips, beetles, birds, and bats (banana), thus representing 37 invertebrate and 7 vertebrate genera (Roubik, 1995; Nabhan & Buchmann, 1997).

Wild pollinators can contribute to crop pollination in four ways. First, they can substitute for the services provided by commercially managed pollinators, replacing them either fully or partially. Second, they can enhance the services provided by managed pollinators through behaviors that increase the effectiveness of the managed pollinator. Third, they can provide services to plants that are not efficiently pollinated by a managed pollinator. Fourth, they can enhance productivity in plants that self-pollinate and for which pollination is consequently rarely managed. In contrast, wild pollinators can also detract from crop pollination in several ways, either by nectar robbery, by competing for pollen with other, potentially superior pollinators, or by transferring heterospecific pollen that clogs stigmas.

When wild bees provide an equivalent (redundant) service to that of the managed pollinator, they can partially or fully substitute for that pollinator. In watermelon production in northern California, honey bees are often imported to fields to provide pollination services. Although their pollination efficiency is low relative to other bee visitors, honey bee contribution to overall pollination is high due to their high abundance under these circumstances. Thirty native bee species also visit watermelon flowers in this area and contribute to pollination. Although none of these species is abundant compared with the artificially high abundances of the honey bee, these species collectively provide on average 28–100% of pollination needs for watermelon (range = 6–100%), depending on the farm environment. Organic farms near natural habitats (low agricultural intensity) reliably receive a large proportion of their pollination requirements from the wild bee community; these farmers never import honey bees to their farms, and the honey bee contribution on its own is not sufficient to provide them with the services they need. Thus such farmers clearly are relying on wild pollinators to some extent. At the other end of the agricultural intensification gradient, conventional farmers far from natural habitat never receive sufficient pollination from wild bees; such farmers always import honey bees to provide pollination services. Nevertheless, they do receive some benefits from wild bee visitors, although they may be unaware of these benefits (Kremen et al., 2002a, 2002b, 2004).

Wild bees can enhance the services provided by managed honey bees via behaviors that increase the rate of pollination. First, they can enhance per-visit pollination efficiency of the honey bee through behavioral interactions. There is a single documented example of this phenomenon (Greenleaf & Kremen, 2006b), but it is likely to be widespread in cropping systems that require movements between cultivars for successful fruit or seed production (e.g., both hybrid seed production systems and many orchard crops). In hybrid sunflower seed production, farmers plant 4 rows of male-sterile, nectar-producing ("female") cultivars for every 6–10 rows of male-fertile, pollen- and nectar-producing ("male") cultivars in a repeating pattern. Honey bees are stocked at 2–2.5 colonies per ha; nonetheless, lack of pollination is a major factor cited by farmers for underproduction. Individual honey bees tend to forage either for pollen or for nectar (Free, 1963). Honey bees had low pollination efficiency on hybrid sunflower relative to the most efficient wild bee visitors (mean of 3 seeds/visit compared with 19). There was a strong linear relationship, however, between per-visit honey bee pollination efficiency and the richness and abundance of wild bees present, increasing the number of seeds set per honey bee visit up to fivefold. Interactions between wild bees and honey bees caused honey bees to transfer more frequently from male to female rows, enhancing their per-visit efficiency. Thus on average, although wild bees contributed only a small proportion of total sunflower pollination directly, they doubled the effectiveness of honey bees and thus the value of the pollination services honey bees provide (Greenleaf & Kremen, 2006b).

Second, better seed and fruit set can result from the combined, complementary foraging activities of honey bees plus wild bees than from that of either group alone. In strawberry, the behavior and morphology of wild bees favors pollination of the basal stigmata, whereas that of honey bees promotes pollination of the apical stigmata. The result of visits by both groups was higher pollination rates (number of fertilized achenes/flower) and larger, more completely formed fruits (Chagnon et al., 1993).

Non-*Apis* bees are more effective pollinators than *Apis mellifera* for some crops that depend on animal pollinators for fruit set, including alfalfa, blueberry, and cranberry (Parker et al., 1987; Delaplane & Mayer, 2000). In these crops, honey bees cannot reliably work the floral mechanism that allows pollination (Proctor et al., 1996). Growers often import large numbers of honey bees, hoping that increasing the frequency of encounters will increase the number of successful pollination events. Alternative pollinators have been domesticated in some cases, including *Megachile rotundata* and *Nomia melanderi* for alfalfa, or *Osmia* species for blueberry, but the use of these pollinators is not widespread (see Crane, 1990, table 8.5). In some cases, growers rely almost entirely on wild bees. In the 1970s in Canada, blueberry growers became acutely aware of their reliance on native pollinators when applications of the insecticide fenitrothion to nearby forests for spruce budworm control greatly reduced many pollinator populations, a reduction that was then correlated with significant crop losses (Kevan & Plowright, 1989).

The majority of economically important fruit and vegetable crops that self-pollinate also benefit from pollination provided by insect vectors by enhanced fruit set and/or size (Klein et al., 2007). The mechanism may be due to increased deposition of self-pollen, cross-pollen, or both, reflecting the contribution of both genetic and physiological factors to fertilization, fruit set, and fruit growth (Proctor et al., 1996; Delaplane & Mayer, 2000). Growers of self-pollinating plants generally do not import pollinators (except in cultivation of greenhouse tomatoes, whose flowers need vibration, either by wind or an insect, to release their pollen); thus enhanced fruit production due to animal-mediated pollination in self-pollinating field crops is generally due to visitation by wild bees (Klein et al., 2003a; Ricketts et al., 2004; Greenleaf & Kremen, 2006a).

Visitation by some insects may actually be detrimental for crop pollination. Insects that cut holes at the base of the flower's corolla in order to obtain nectar resources may reduce a flower's attractiveness and deter other insects from visiting and pollinating the plant (Irwin et al., 2001). Insects that visit multiple flowering species may transfer heterospecific pollen during visits to crop flowers, which could then clog stigmas, reducing both the effectiveness of that visit and of subsequent visits by the same or other pollinators. In general, non-*Apis* individuals are thought to exhibit lower flower constancy than honey bees (Slaa & Biesmeijer, 2005); thus it is conceivable that non-*Apis* wild pollinators could reduce pollination services provided by honey bees through stigma clogging, although I know of no examples in crops.

Insects (usually bees, but also pollen-eating beetles) that remove large amounts of pollen while depositing only tiny amounts can be negative, rather than positive, for pollination function in crops. The extent to which a given species (whether wild or managed) is detrimental versus beneficial for crop pollination services depends on three things: (1) its species-specific behavior, leading to its mean ratio of pollen removal to deposition; (2) the composition of the pollinator community; and (3) whether the amount of available pollen is a limiting factor. Under limiting conditions (i.e., all pollen produced is removed), if one visiting species has a high ratio of pollen removal to deposition relative to other community members, its contribution to pollination will be negative, because it removes pollen from the system that other pollinators could otherwise deposit. If it has a low removal to deposition ratio relative to other species, or if there are no other pollinating species, then it increases pollination (Thomson & Thomson, 1992; Thomson & Goodell, 2001). If the amount of pollen is not limiting, however, then more visits from any visitor that deposits any amount of pollen add to the total pollen deposited on the crop. Pollen supply will depend greatly on the cultivar, crop breeding system, and other details of cultivation (e.g., proportion of plants supplying pollen in the crop field).

Although wild bee pollinators may augment or in some cases substitute for the services provided by commercially managed pollinators, it is important to recognize some inherent limitations to services provided by wild, unmanaged bees. Wild pollinator populations are notably variable in space and time (Roubik, 2001; Williams et al., 2001); thus services they provide may not be consistent enough to meet the needs of large-scale intensive agriculture. Unlike the honey bee, which forms permanent colonies of 30,000 to 50,000 individuals, non-*Apis* bees often have relatively small population sizes, particularly at the beginning of the flight season for multivoltine or social species with multiple generations of workers within a season.

Commercially managed pollinators are clearly critical to the success of modern agriculture, but wild, unmanaged pollinators, despite the caveats noted previously, could reduce the risk of depending overly on just one or a few commercial species. Risks from relying on only a few species come from: (1) the challenges of maintaining a stable supply of commercial pollinators, given problems of managing the genetics, pathogens, and parasites of honey bees and other commercial pollinators (National Research Council, 2007); and (2) limitations in pollination services provided by only a few pollinator species (see the upcoming section on the role of diversity). For example, honey bee workers communicate with each other about the spatial location and quality of foraging resources. This social behavior can lead to massive recruitment of workers to a crop that is rewarding in pollen and nectar, but it may also result in workers concentrating in selected areas of the field, which can bring about uneven crop pollination across the field. In the worst case, honey bee workers leave the crop altogether to forage on more attractive noncrop resources (Free, 1968). Although less numerous and certainly more patchy in their distributions, wild bees may complement the services provided by honey bees (Chagnon et al., 1993) and spread pollinators over a larger area of the crop (Proctor et al., 1996). Given their small, patchy populations, however, the goal of managing for wild pollinators should be to augment the services provided by commercial pollinators by maintaining diverse communities that collectively provide more stable services than any individual wild pollinator species could (Tilman et al., 1998; Klein et al., 2003b; Kremen et al., 2002b, 2004).

Economic Value of Services From Wild Pollinators

Estimating the economic value of services provided by wild pollinators is complicated for three reasons. First, different approaches to estimating the value of pollination services yield widely differing results (Kremen et al., 2007). The lowest value would be the cost to replace wild bee pollination services with commercial pollinators (Muth & Thurman, 1995). The highest value comes from establishing the proportional dependence of a crop on animal pollination and then multiplying this proportional dependence by the gross value of the crop produced (Robinson et al., 1989a, 1989b). Second, in situations in which both managed and wild bees contribute to pollination services, determining the contribution of each requires intensive field documentation (Greenleaf & Kremen, 2006b; Olschewski et al., 2006; Priesset al., 2007). Such information is rarely available. Nabhan and Buchmann (1997) have suggested that contributions from wild bees would be similar to those from managed bees, but using the same basic approach of Robinson et al. (1989a, 1989b), Losey and Vaughan (2006) estimated the contribution to U.S. fruit and vegetable production of wild bees at $3.07 billion, less than 20% of the contribution of honey bees ($17.01 billion). Third, interactions between wild bees and honey bees that augment pollination services require yet another level of field documentation (e.g., Chagnon et al., 1993; Greenleaf & Kremen, 2006b) and may dramatically increase the value attributed to wild bees. For example, in the hybrid sunflower seed production described earlier, Greenleaf and Kremen (2006b) attributed only 7.3% of the gross value of the U.S. hybrid sunflower seed crop ($26.1 million) to wild bees through direct pollination but an additional 39.8% to their enhancement of the pollination services provided by honey bees. The direct contribution provided by honey bees without the beneficial effects of wild bees was 52.9%.

Effects of Agricultural Land Use on Wild Bee Communities and Pollination Services to Crops

Agricultural land use may have either positive or negative effects on pollinator communities and the services they provide, depending on the intensity of agricultural land use, the spatial scale (Tscharntke et al., 2005), and the biome, although too few studies have been conducted to predict these effects with certainty. Both site and landscape-scale factors may be important (see figure 2.1). In a Mediterranean biome in California, agricultural intensification, which included both the reduction of nearby natural habitat and the predominance of large-scale industrialized agriculture (for a definition, see Tscharntke et al., 2005), led to reductions in the species richness and abundance of wild bee pollinators on watermelon (Kremen et al., 2002b, 2004), tomato (Greenleaf & Kremen, 2006a), and sunflower (Greenleaf & Kremen, 2006b), with concomitant estimated reductions in the services wild bees provide to these crops. In these studies, a common factor influencing wild bee distributions appeared to be the area of nearby natural habitats (chaparral and oak woodlands) within several kilometers of the farm site. The proportional area or proximity of natural habitat was positively correlated with bee species richness, abundance, the number of nesting bees found on farms, and the magnitude and stability of pollination services provided by wild bees (Kremen et al., 2004; Greenleaf & Kremen 2006a, 2006b; Kim et al., 2006). Local farm management type (organic vs. conventional) only weakly affected these community response variables once the landscape level effects were factored out, although for sunflower, the interannual continuity of sunflower availability within foraging range of bees was equally important (Greenleaf & Kremen, 2006b). Individual bee species were differentially sensitive to the gradient of agricultural intensification, but none increased in response to it (Kremen, 2004). The species that were the more effective pollinators in watermelon were also the more sensitive to agricultural intensification; thus their loss exacerbated the effects on pollination services (Larsen et al., 2005).

Similarly, in the neotropics, distance to wild forest patches significantly influenced the richness and abundance of wild bees visiting and pollinating coffee in Costa Rica (Ricketts, 2004) and grapefruit in Argentina (Chacoff & Aizen, 2006). These wild bees included indigenous solitary and social bees and feral colonies of introduced *Apis mellifera scutellata*. Over a span of 100 meters from the forest edge, visitation dropped precipitously by 75% (Ricketts et al., 2004), although a decline in pollination services was

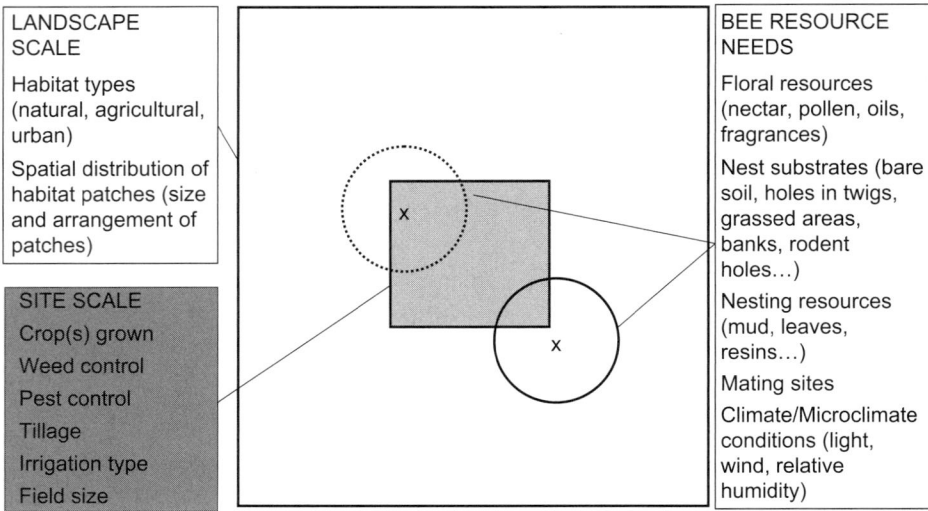

Figure 2.1 Schematic depicting the influence of site and landscape-level factors on bees in an agricultural landscape. The small gray box denotes a farm field embedded in a larger landscape, which generally includes multiple habitat types. Each x denotes the nesting site of one female bee, and the circle around each x denotes the foraging range of the bee. One female nests off the farm (solid circle) and the other nests on the farm (dashed circle), but in both cases, their foraging ranges encompass both farm and off-farm areas. Bees require floral and nesting resources that are available within their foraging range and throughout their adult flight period, as well as suitable climatic/microclimatic conditions for flying, foraging, and mating. At the farm scale, management practices influence both the availability of floral and nesting resources and microclimate conditions through the choice of crops and cultural practices, including weed control, irrigation, and tillage. Use of pesticides influences mortality rates of bees and bee predators and parasitoids. At the landscape scale, the heterogeneity of the habitat influences the diversity and abundance of available floral resources and of nesting sites/substrates within the bees' foraging range (circle), which tends to be larger in species of larger size.

not observed until 1600 meters (Ricketts, 2004). Similarly, richness and visit frequency of native wild bees visiting grapefruit (*Citrus paradisi*) declined precipitously with distance from the forest edge in Argentina (by eightfold within 1,000 meters), and the visit frequency of feral *Apis mellifera scutellata*, which accounted for 95% of visits, dropped by twofold over the same distance (Chacoff & Aizen, 2006). In coffee fields in Indonesia, distance to wild habitat affected the richness of native social but not solitary bees, whereas light levels within the fields were strongly, positively correlated with solitary bee richness and with the abundance of both solitary and social bees. Fruit set was significantly correlated with both factors (Klein et al., 2003b). In macadamia (*Macadamia integrifolia*) orchards in southern Queensland and New South Wales, Australia, the abundance of its most common native pollinator, *Trigona carbonaria*, but not of managed *Apis mellifera*, correlated with the proportional area of *Eucalyptus* forests within 1 kilometer of orchards

(Heard & Exley, 1994). In contrast, on the Atherton Tablelands in northern Queensland, where honey bees were not managed for pollination, distance from rainforest corresponded with a decline in both feral *Apis mellifera* visits and in fruit set of macadamia, although there was no correlation between fruit set per raceme and *A. mellifera* abundance per site (Blanche et al., 2006). In the same area, beetle visitors to custard apple (*Annona squamosa* × *A. cherimola*) declined in diversity and abundance with distance from rainforest habitat, with a corresponding decline in fruit production (Blanche & Cunningham, 2005; Pritchard, 2005). Similarly, wild stingless bees (*Trigona* spp.), but not feral *A. mellifera*, declined in abundance with increasing distance from rainforest in longan orchards (*Dimocarpus longan*), with a corresponding decrease in fruit set (Blanche et al., 2006). In these mosaic environments of tropical forest and agriculture, forest patches again appear to play an important role in providing habitat for native and non-native bee pollinators of crops and thus for pollination services.

In temperate agricultural landscapes in Europe with patches of seminatural habitats (calcareous grasslands, woods, meadows, and other habitats), distance to these patches influenced diversity, abundance and pollination services provided by social and solitary bees. In both of two self-incompatible plants, mustard (*Sinapsis arvensis*) and radish (*Raphanus sativus*), reproductive output was halved first at a 250-meter distance from patches and then again at 1,000 meters (Steffan-Dewenter & Tscharntke, 1999). In contrast, the abundance of common *Bombus* species in this same landscape type did not correlate with the proportional area of seminatural habitat but did positively correlate with the proportional area of such mass-flowering crops as oilseed rape, clover, and sunflower. This finding suggests that the enormous flush of pollen and nectar resources provided by large fields of monoculture crops can promote abundance of selected bee species (Westphal et al., 2003), particularly if pollen and nectar resources provided by these crops are staggered across the bumble bee flight season.

Although wild bees on crops generally show a decline of diversity, abundance, and services with agricultural intensification (*sensu* Tscharntke et al., 2005), not all studies of bee communities (including bees visiting noncrop resources) show the same diversity and abundance trends. For example, in the Atlantic Coastal Pine Barren's ecoregion of the northeastern United States, the richness and abundance of bee species in fragments of this habitat increased significantly when surrounded by a predominantly agricultural matrix compared with a predominantly forested matrix (Winfree et al., 2007). Agricultural habitats also had significantly greater richness and abundance than naturally forested habitats, and more species were found to be unique to the agricultural areas as compared with the forested areas. Forests in the Pine Barrens are composed of a pine overstory with a low-diversity, ericaceous understory. Both floral richness and abundance were higher in agricultural areas than within Pine Barren forests. In this system, agriculture apparently enhances rather than detracts from bee richness and abundance, although it must be noted that the intensity of agricultural land use is relatively low (approximately 30% of land uses within 1.6 km of sample sites) compared with other study systems. In this case, agriculture may mimic various early successional habitats in which bee species often thrive (e.g., Carvell, 2002; Potts et al., 2003; Grixti & Packer, 2006). Positive effects of agriculture on pollinator communities may be more likely to occur in regions in which the presence of agriculture increases rather than decreases habitat heterogeneity within the foraging range of bees (e.g., <2 km), such as farming landscapes that include relatively small field sizes, mixed crop types within or between fields, and patches of noncrop vegetation, such as hedgerows, fallow fields, meadows, and seminatural wood or shrublands (Eltz et al., 2002; Tscharntke et al., 2005).

In summary, based on reported studies, pollination services provided by wild bees are most likely being reduced in many of the areas in which they could be contributing to crop pollination. At the same time, numbers of commercially managed honey bee colonies have also declined, and challenges for honey bee management are increasing (National Research Council, 2007). Yet there are comparatively few documented instances of shortages in pollination services. This suggests that we are not yet in crisis; but, to be cautious, we should take preventive measures now. In particular, our heavy reliance on honey bees makes production of some crops (especially almond and other orchard crops) vulnerable to sudden, unforeseen changes in its abundance, such as appear to be occurring with increasing frequency in the United States following winter season declines (National Research Council, 2007).

Role of Diversity

A more diverse community of wild pollinators can provide a greater amount of pollination services to a greater number of crops with greater stability. More diverse communities of pollinators in agricultural systems also have greater total abundances and rates of visitation to crop flowers (Steffan-Dewenter & Tscharntke, 1999; Klein et al., 2003b; Ricketts, 2004; Larsen et al., 2005; Pritchard, 2005; Chacoff & Aizen, 2006). The strikingly consistent positive relationship between abundance and richness across these studies suggests that the loss of richness will generally reduce the number of visits and hence the level of pollination services provided to crops by the wild bee community, given a strong correlation between visit number and pollination services across systems (Vázquez et al., 2005).

Although many pollinator species that visit crops are generalists, different crop species nonetheless attract different, albeit partially overlapping, sets of pollinator species from the local species pool. Therefore, maintaining diverse pollinator communities locally is important for providing pollination services to a more diverse set of crops. Within a crop, a diverse group of pollinator species can provide better pollination services than a single species can, due to different foraging behaviors (e.g., strawberry; Chagnon et al., 1993) or to interactions that influence foraging movements (e.g., sunflower; Greenleaf & Kremen, 2006b). Within a crop, diversity of the pollinator community is also important for ensuring the stability of pollination services across time and space. Several lines of reasoning support this assertion. From theoretical principles, we know that more diverse communities whose populations fluctuate in a random, uncorrelated fashion will provide more consistent services than will less diverse communities (the portfolio effect; Tilman et al., 1998). Empirical work supports the claims of theory, although few studies have yet been conducted. Richer communities provided more stable pollination services to watermelon crops from day to day within a season (Kremen, unpublished data) and from bush to bush within a coffee field (Klein et al., 2003b; Steffan-Dewenter et al., 2006). High-diversity communities may include an array of species with broader physiological and behavioral ranges that are able to fly and to pollinate flowers under a wider array of environmental circumstances and thus to provide greater consistency than lower-diversity communities (Herrera, 1995; Bishop & Armbruster, 1999; Klein et al., 2003b).

Insect populations, especially bees, fluctuate greatly in the wild from year to year, as well as within seasons and across space (Herrera, 1988; Wolda, 1988; Roubik, 2001; Williams et al., 2001). Such transient losses are unlikely to affect pollination services to a given plant species as long as the system is relatively diverse (Williams et al., 2001; Memmott et al., 2004; Morris, 2003). In the watermelon system, entirely different bee species predominated in their visit frequencies (abundances) in 2 successive years, and hence their species-specific contributions to watermelon pollination. In both years, however, the community collectively provided sufficient services on high-diversity farms (Kremen et al., 2002b). In Costa Rica, decline in 1 year in the abundance of feral non-native *A. mellifera scutellata* was partially balanced by increases in abundances of native species (Ricketts, 2004). In these systems, managing for wild bee richness, rather than for the abundance of a particular species, is an important factor in maintaining a consistent level of service.

Managing for Wild Bee Populations and Services in the Agroecosystem

Wild pollinators are mobile organisms that often utilize a multiplicity of resources; often, different resources are localized in different, noncontiguous habitats (Westrich, 1996; see figure 2.1). Maintaining wild pollinator populations, therefore, requires understanding resource requirements and then managing habitats and landscapes to provide food resources, nesting habitats, overwintering habitats, and breeding areas. Resources must be available within foraging/dispersal distances, or organisms will die or have low reproductive rates. Managing for pollinator populations requires thinking not only at the site scale but also at the landscape and even the regional scale.

For example, in California, many of the bee species visiting crop plants are generalists with long flight periods (e.g., *Bombus* and *Halictus* species; Kremen et al., 2002a). They require a suite of floral resources stretching from early spring to mid-fall (January to October for some species) and may, therefore, depend not only on the weed and crop resources that are available on farm lands but also on wild, native plants that occur in neighboring riparian, chaparral, and oak woodland areas (Kremen et al., 2002a; Williams & Kremen, 2007). Decreasing the area of natural habitat within a given radius of a farm site (or, conversely, increasing the distance to patches of natural habitat; Harrison & Fahrig, 1995) could increase the energetic requirements to obtain floral resources and thus to produce offspring (Orians & Pearson, 1979). Species that depend on native plants for all or part of their life cycle may either drop out completely or diminish in abundance on farms with little natural habitat within foraging range of the bee species (e.g., Larsen et al., 2005). Conversely, alternative resources provided in the agroecosystem may mitigate the loss of resources in natural habitat. Williams and Kremen (2007) monitored offspring production in experimental cavity nests of *Osmia lignaria*, the blue orchard bee, and found that resources available on organic farms partially substituted for preferred resources from wild plants, reducing the dependence of *O. lignaria* productivity on the proximity of natural habitat on organic farms. In contrast, on conventional farms that did not have such onsite resources, *Osmia* productivity rates declined significantly with increasing distance to natural habitat, and offspring survival was below replacement in the most isolated sites.

Bee species will differ in their capacity to nest on farms. Some bee species require rodent nests or cavities in wood in which to nest, and these may not be available on farm sites. Bees that excavate nests in the ground may suffer mortality from flood irrigation and plowing if they nest in agricultural fields (Shuler et al., 2005). In California, less than half of the ground-nesting bee species found visiting sunflower were also found nesting on or in sunflower fields (Kim et al., 2006). If nests are located offsite, it provides a constraint to the distance they will be able to forage; thus only farms within foraging range will receive pollination services. Foraging ranges differ widely among species and are strongly related to body size; in the California system foraging ranges are below 2 kilometers (Greenleaf et al., 2007).

It therefore seems clear that managing at the landscape scale, as well as at the site level, will be important for restoring, preserving, or maintaining pollinator communities and services. How much land is enough to provide sustainable pollinator communities and services? Only a few studies have addressed this issue, and far more work remains to be done. Kremen et al. (2004) observed a log-linear relationship between the amount of pollination services provided to watermelon and the proportional area of natural habitat within several kilometers of the farm site. Full pollination services could be provided by wild bee communities at approximately 30% natural habitat cover or above. We know even less about how patches of habitat should be optimally configured to deliver pollination services to crops in the surrounding agricultural matrix. Indeed, because many pollinator populations are not limited to natural or seminatural habitat patches but, rather, utilize different elements in both natural and agricultural areas, a better question may be, How complex should the landscape be to ensure population persistence of pollinators (Tscharntke et al. 2005)? In addition, we know little about the factors that limit bee populations. Is it floral resources, nesting sites, or both? What role do predators, parasitoids, parasites, and disease play in limiting bee populations, and how do these factors respond to landscape structure?

Although our knowledge is incomplete, a great deal could currently be done to improve the situation for wild pollinators in agricultural landscapes, acting both at the site (field) and landscape scale. Site-level management actions could include introducing multicropping, allowing cover crops to flower or permitting weedy borders, restoring native plant hedgerows that consist of phenological suites of plants that support pollinators, creating small patches of bare ground for nesting, installing bumble bee boxes and trap nests, and leaving small patches of woods for cavity-nesting bees (Vaughan et al., 2004). In areas of Eastern Europe, alfalfa growers successfully managed for a wild pollinator, the alfalfa gray-haired bee (*Rhophitoides canus*), by carefully timing and spacing cutting of alfalfa bloom so as to provide alfalfa bloom throughout the life cycle and bare soil for nesting during the peak nesting period (Bosch, 2005). In the United Kingdom, growers plant flower-rich field margins to enhance pollinator abundance on farms (Dover, 1997; Carvell et al., 2004). The specific composition of plantings may be important in determining pollinator abundance and diversity (Gurr et al., 2004; Pywell et al., 2005), but it is not known whether such field strips enhance population size and persistence of pollinators or simply redistribute them within the landscape.

Such small-scale restorations or changes in field management practices could promote floral and nesting resources for bees at little or no cost to farmers. To the extent that these practices enhance populations rather than redistribute individuals, small-scale changes that initially incur small annual costs could have cumulative effects that ultimately pay for themselves, allowing farmers to reduce rental payments for honey bees or to weather periods of scarcity of commercially managed pollinators (Kremen et al., 2002b). Such management practices could ultimately transform farm sites from sinks to sources of native bees by increasing reproductive rates above replacement (*sensu* Pulliam, 1988). In California, organic farms acted like source habitats for experimental *O. lignaria* populations compared with conventional farms, which acted like sinks, with reproductive rates above replacement on organic farms but below replacement on conventional ones (Williams & Kremen, 2007). Landscape-level management actions could include coordinating small-scale efforts among growers to build larger "patches" of bee-friendly farms and enhance connectivity between them through restoration of riparian or other corridors and through conservation of existing seminatural and natural habitats. Such actions are evidently much more difficult and expensive to implement and will generally be more likely to happen if they simultaneously promote multiple ecosystem service benefits (Balvanera et al., 2001). It may rarely be the case that the economic benefits from enhanced pollination services alone are sufficient to bear the costs of managing sites and landscapes for wild pollinators (Olschewski et al., 2006).

References

Balvanera, P., Daily, G. C., Ehrlich, P. R., Ricketts, T. H., Bailey, S. A., Kark, S., et al. (2001). Conserving biodiversity and ecosystem services: Conflict or reinforcement? *Science*, 291, 2047.

Batra, S. W. T. (2001). Coaxing pollen bees to work for us. In C. Stubbs & F. Drummond (Eds.), *Bees and crop pollination: Crisis, crossroads, conservation* (85–93). Lanham, MD: Entomological Society of America.

Beetsma, J. (1994). The *Varroa* mite: A devastating parasite of western honey bees and an economic threat to beekeeping. *Outlook on Agriculture*, 23, 169–175.

Bishop, J. A., & Armbruster, W. S. (1999). Thermoregulatory abilities of Alaskan bees: Effects of size, phylogeny and ecology. *Functional Ecology, 13,* 711–724.

Blanche, R., & Cunningham, S. A. (2005). Rain forest provides pollinating beetles for atemoya crops. *Journal of Economic Entomology, 98,* 1193–1201.

Blanche, R., Ludwig, J. A., & Cunningham, S. A. (2006). Proximity of rainforest enhances pollination and fruit set orchards. *Journal of Applied Ecology, 43,* 1182–1187.

Bosch, J. (2005). The contribution of solitary bees to crop pollination: From ecosystem service to pollinator management. In J. M. Guerra-Sanz, A. Roldán, & A. Mena Granero (Eds.), *First short course on pollination of horticulture plants* (151–165). Almería, Spain: CIFA La Mojonera.

Carvell, C. (2002). Habitat use and conservation of bumble bees (*Bombus* spp.) under different grassland management regimes. *Biological Conservation, 103,* 33–49.

Carvell, C., Meek, W. R., Pywell, R. P., & Nowakowski, M. (2004). The response of foraging bumble bees to successional changes in newly created arable field margins. *Biological Conservation, 118,* 327–339.

Chacoff, N., & Aizen, M. (2006). Edge effects on flower-visiting insects in grapefruit plantations bordering premontane subtropical forest. *Journal of Applied Ecology, 43,* 18–27.

Chagnon, M., Gingras, J., & Deoliveira, D. (1993). Complementary aspects of strawberry pollination by honey and indigenous bees (Hymenoptera). *Journal of Economic Entomology, 86,* 416–420.

Crane, E. (1990). *Bees and beekeeping: Science, practice and world resources.* Ithaca, NY: Comstock.

Delaplane, K. S., & Mayer, D. F. (2000). *Crop pollination by bees.* New York: CABI.

Dover, J. W. (1997). Conservation headlands: Effects on butterfly distribution and behaviour. *Agriculture Ecosystems and Environment, 63,* 31–49.

Eltz, T., Bruhl, C. A., van der Kaars, S., & Linsenmair, K. E. (2002). Determinants of stingless bee nest density in lowland dipterocarp forests of Sabah, Malaysia. *Oecologia, 131,* 27–34.

Elzen, G. W., & Hardee, D. D. (2003). United States Department of Agriculture–Agricultural Research Service research on managing insect resistance to insecticides. *Pest Management Science, 59,* 770–776.

Free, J. B. (1963). The flower constancy of honey bees. *Journal of Animal Ecology, 32,* 119–131.

———. (1968). Dandelion as a competitor to fruit trees for bee visits. *Journal of Applied Ecology, 5,* 169–178.

———. (1993). *Insect pollination of crops.* San Diego, CA: Academic Press.

Greenleaf, S., & Kremen, C. (2006a). Wild bee species increase tomato production but respond differently to surrounding land use in Northern California. *Biological Conservation, 133,* 81–87.

Greenleaf, S. S., & Kremen, C. (2006b). Wild bees enhance honey bees' pollination of hybrid sunflower. *Proceedings of the National Academy of Sciences of the USA, 103,* 13890–13895.

Greenleaf, S., Williams, N., Winfree, R., & Kremen, C. (2007). Bee foraging ranges and their relationships to body size. *Oecologia, 153,* 589–596.

Griffiths, D. (1986). Summary of the present status of *varroatosis* in Europe. In R. Cavalloro (Ed.), *European research on varroatosis control* (11–13). Rotterdam, Netherlands: Balkema.

Grixti, J. C., & Packer, L. (2006). Changes in the bee fauna (*Hymenoptera: Apoidea*) of an old field site in southern Ontario, revisited after 34 years. *Canadian Entomologist, 138,* 147–164.

Gurr, G. M., Wratten, S. D., & Altieri, M. (2004). *Ecological engineering for pest management: Advances in habitat manipulation for pest management.* Melbourne, Australia: CSIRO.

Harrison, S., & Fahrig, L. (1995). Landscape pattern and population conservation. In L. Hansson, L. Fahrig, & G. Merriam (Eds.), *Mosaic landscapes and ecological processes* (293–308). London: Chapman & Hall.

Heard, T. A., & Exley, E. M. (1994). Diversity, abundance, and distribution of insect visitors to macadamia flowers. *Environmental Entomology, 23,* 91–100.

Herrera, C. M. (1988). Variation in mutualisms: The spatio-temporal mosaic of a pollinator assemblage. *Biological Journal of the Linnean Society, 35,* 95–125.

———. (1995). Microclimate and individual variation in pollinators: Flowering plants are more than their flowers. *Ecology, 76,* 1516–1524.

Irwin, R. E., Brody, A. K., & Waser, N. W. (2001). The impact of flower larceny on individuals, populations, and communities. *Oecologia, 129,* 161–168.

Kevan, P. G., Clark, E. A.,& Thomas, V. G. (1990). Insect pollinators and sustainable agriculture. *American Journal of Alternative Agriculture, 5,* 12–22.

Kevan, P. G., & Phillips, T. P. (2001). The economic impacts of pollinator declines: An approach to assessing the consequences. *Conservation Ecology, 5,* 8.

Kevan, P. G., & Plowright, R. C. (1989). Fenitrothion and insect pollinators. In W. R. Ernst, P. A. Pearce, & T. L. Pollock (Eds.), *Environmental effects of fenitrothion use in forestry: Impacts on insect pollinators, songbirds, and aquatic organisms* (13–42). Dartmouth, Nova Scotia: Environment Canada, Conservation and Protection, Atlantic Branch.

Kim, J., Williams, N., & Kremen, C. (2006). Effects of cultivation and proximity to natural habitat on ground-nesting native bees in California sunflower fields. *Journal of the Kansas Entomological Society, 79,* 309–320.

Klein, A. M., Steffan-Dewenter, I., & Tscharntke, T. (2003a). Bee pollination and fruit set of *Coffea arabica* and *C. canephora* (Rubiaceae). *American Journal of Botany, 90,* 153–157.

———. (2003b). Fruit set of highland coffee increases with the diversity of pollinating bees. *Proceedings of the Royal Society of London: Series B. Biological Sciences, 270,* 955–961.

Klein, A. M., Vaissière, B., Cane, J. H., Steffan-Dewenter, I., Cunningham, S. A., Kremen, C., et al. (2007). Importance of crop pollinators in changing landscapes for world crops. *Proceedings of the Royal Society of London: Series B. Biological Sciences, 274,* 303–313.

Komeili, A. B. (1988). The impact of the *Varroa* mite on Iranian commercial beekeeping. *American Bee Journal, 128,* 423–424.

Kremen, C. (2004). Pollination services and community composition: Does it depend on diversity, abundance, biomass, or species traits? In B. M. Freitas & J. O. P. Pereira (Eds.), *Solitary bees: Conservation, rearing and management for pollination* (115–124). Ceara, Brazil: University Dederal do Ceara.

Kremen, C., Bugg, R. L., Nicola, N., Smith, S. A., Thorp, R. W., & Williams, N. M. (2002a). Native bees, native plants and crop pollination in California. *Fremontia, 30,* 41–49.

Kremen, C., Williams, N. M., Aizen, M. A., Gemmill-Harren, B., LeBuhn, G., Minckley, R., et al. (2007). Pollination and other ecosystem services produced by mobile organisms: A conceptual framework for the effects of land-use change. *Ecology Letters, 10,* 299–314.

Kremen, C., Williams, N. M., Bugg, R. L., Fay, J. P., & Thorp, R. W. (2004). The area requirements of an ecosystem service: Crop pollination by native bee communities in California. *Ecology Letters, 7,* 1109–1119.

Kremen, C., Williams, N. M., & Thorp, R. W. (2002b). Crop pollination from native bees at risk from agricultural intensification. *Proceedings of the National Academy of Sciences of the USA, 99,* 16812–16816.

Larsen, T. H., Williams, N., & Kremen, C. (2005). Extinction order and altered community structure rapidly disrupt ecosystem functioning. *Ecology Letters, 8,* 538–547.

Losey, J. E., & Vaughan, M. (2006). The economic value of ecological services provided by insects. *Bioscience, 56,* 311–323.

Memmott, J., Waser, N. M., Price, M. V. (2004). Tolerance of pollination networks to species extinctions. *Proceedings of the Royal Society of London: Series B. Biological Sciences, 271,* 2605–2611.

Morandin, L. A., & Winston, M. L. (2005). Wild bee abundance and seed production in conventional, organic, and genetically modified canola. *Ecological Applications, 15,* 871–881.

Morris, W. (2003). Which mutualists are most essential? Buffering of plant reproduction against the extinction of pollinators. In P. Kareiva & S. A. Levin (Eds.), *The importance of species: Perspectives on expendability and triage* (260–280). Princeton, NJ: Princeton University Press.

Morse, R. A., & Goncalves, L. S. (1979). *Varroa* disease: A threat to world beekeeping. *Bee Culture, 107,* 179–181.

Muth, M. K., & Thurman. W. N. (1995). Why support the price of honey? *Choices, 10,* 19–22.

Nabhan, G. P., & Buchmann, S. (1997). Services provided by pollinators. In G. C. Daily (Ed.), *Nature's services: Societal dependence on natural ecosystems* (133–150). Washington, DC: Island Press.

National Research Council. (2007). *Status of pollinators in North America.* Washington, DC: National Academies Press.

Olschewski, R., Tscharntke, T., Benítez, P. C., Schwarze, S., & Klein, A. (2006). Economic evaluation of pollination services comparing coffee landscapes in Ecuador and Indonesia. *Ecology and Society, 11.* Retrieved 2007 from http://www.ecologyandsociety.org/vol11/iss11/art17/.

Orians, G., & Pearson, N. (1979). On the theory of central place foraging. In D. J. Horn, B. R. Stairs, & R. D. Mitchell (Eds.), *Analysis of ecological systems* (155–177). Columbus: Ohio State University Press.

Parker, F. D., Batra, S. W. T., & Tepedino, V. J. (1987). New pollinators for our crops. *Agricultural Zoology Reviews, 2,* 279–304.

Potts, S. G., Vulliamy, B., Dafni, A., O'Toole, C., Roberts, S., & Willmer, P. (2003). Response of plant-pollinator communities following fire: Changes in diversity, abundance and reward structure. *Oikos, 101,* 103–112.

Priess, J., Mimler, M., Klein, A., Schwarze, S., Tscharntke, T., & Steffan-Dewenter, I. (2007). Linking deforestation scenarios to pollination services and economic returns in coffee agroforestry systems. *Ecological Applications, 17,* 407–417.

Pritchard, K. (2005). The unseen costs of agricultural expansion across a rainforest landscape: Depauperate pollinator communities and reduced yield in isolated crops. Unpublished master's thesis, James Cook University, Queensland, Australia.

Proctor, M., Yeo, P., & Lack, A. (1996). *The natural history of pollination.* Portland, OR: Timber Press.

Pulliam, H. R. (1988). Sources, sinks, and population regulation. *American Naturalist, 132,* 652–661.

Pywell, R. F., Warman, E. A., Carvell, C., Sparks, T. H., Dicks, L. V., Bennett, D., et al. (2005). Providing foraging resources for bumble bees in intensively farmed landscapes. *Biological Conservation, 121,* 479–494.

Ricketts, T. H. (2004). Tropical forest fragments enhance pollinator activity in nearby coffee crops. *Conservation Biology, 18,* 1262–1271.

Ricketts, T. H., Daily, G. C., Ehrlich, P. R., & Michener, C. D. (2004). Economic value of tropical forest to coffee production. *Proceedings of the National Academy of Sciences of the USA, 101,* 12579–12582.

Robinson, W. S., Nowogrodzki, R., & Morse, R. A. (1989a). The value of honey bees as pollinators of U.S. crops. *American Bee Journal, 129,* 477–487.

———. (1989b). The value of honey bees as pollinators of U.S. crops. *American Bee Journal, 129,* 411–423.

Roubik, D. (1995). *Pollination of cultivated plants in the tropics.* Rome: Food and Agriculture Organization.

———. (2001). Ups and downs in pollinator populations: When is there a decline? *Conservation Ecology, 5,* 2.

Shuler, R. E., Roulston, T. H., & Farris, G. E. (2005). Farming practices influence wild pollinator populations on squash and pumpkin. *Journal of Economic Entomology, 98,* 790–795.

Slaa, J., & Biesmeijer, K. (2005). Flower constancy. In A. Dafni, P. G. Kevan, & B. C. Husband (Eds.), *Practical pollination biology* (381–400). Cambridge, Ontario, Canada: Enviroquest.

Steffan-Dewenter, I., & Tscharntke, T. (1999). Effects of habitat isolation on pollinator communities and seed set. *Oecologia, 121,* 432–440.

Steffan-Dewenter, I., Klein, A.-M., Gaebele, V., Alfert, T., & Tscharntke, T. (2006). Bee diversity and plant-pollinator interactions in fragmented landscapes. In N. M. Waser & J. Ollerton (Eds.), *Specialization and generalization in plant-pollinator interactions* (387–410). Chicago: University of Chicago Press.

Thomson, J. D., & Goodell, K. (2001). Pollen removal and deposition by honeybee and bumblebee visitors to apple and almond flowers. *Journal of Applied Ecology, 38,* 1032–1044.

Thomson, J. D., & Thomson, B. A. (1992). Pollen presentation and viability schedules in animal-pollinated plants: Consequences for reproductive success. In R. Wyatt (Ed.), *Ecology and evolution of plant reproduction: New approaches* (1–24). New York: Chapman & Hall.

Tilman, D., Lehman, C. L., & Bristow, C. E. (1998). Diversity-stability relationships: Statistical inevitability or ecological consequence? *American Naturalist, 151,* 277–282.

Tscharntke, T., Klein, A. M., Kruess, A., Steffan-Dewenter, I., & Thies, C. (2005). Landscape perspectives on agricultural intensification and biodiversity: Ecosystem service management. *Ecology Letters, 8,* 857–874.

Vaughan M., Shepard, M., Kremen, C., & Black, S. H. (2004). *Farming for bees: Guidelines for providing native bee habitat on farms.* Portland, OR: The Xerces Society.

Vázquez, D. P., Morris, W. F., & Jordano, P. (2005). Interaction frequency as a surrogate for the total effect of animal mutualists on plants. *Ecology Letters, 8,* 1088–1094.

Westphal, C., Steffan-Dewenter, I., & Tscharntke, T. (2003). Mass flowering crops enhance pollinator densities at a landscape scale. *Ecology Letters, 6,* 961–965.

Westrich, P. (1996). Habitat requirements of central European bees and the problems of partial habitats. In A. Matheson, S. L. Buchmann, C. O'Toole, P. Westrich, & I. H. Williams (Eds.), *The conservation of bees* (1–16). London: Academic Press.

Williams, N., & Kremen, C. (2007). Floral resource distribution among habitats determines productivity of a solitary bee, *Osmia lignaria*, in a mosaic agricultural landscape. *Ecological Applications, 17,* 910–921.

Williams, N. M., Minckley, R. L., & Silveira, F. A. (2001). Variation in native bee faunas and its implications for detecting community changes. *Conservation Ecology, 5,* 7.

Winfree, R., Griswold, T., & Kremen, C. (2007). Effect of human disturbance on bee communities in a forested ecosystem. *Conservation Biology, 21,* 213–223.

Wolda, H. (1988). Insect seasonality: Why? *Annual Review of Ecology and Systematics, 19,* 1–18.

3   Crop Pollination in Greenhouses

José M. Guerra-Sanz

Introduction

Greenhouse vegetable cultivation is a production system providing a high income per unit area due to various benefits, such as year round production, improvements in crop quality, and increases in yield. Greenhouses allow a more efficient use of water, fertilizers, pesticides, and labor. For all these reasons, protected cultivation, especially greenhouse production, has increased worldwide over the past three decades. Total protected cultivation acreage in the world has already reached 2 million ha, and total greenhouse acreage is over 700,000 ha (Pardossi et al., 2004). The Mediterranean Basin is one area of concentrated greenhouse production. Approximately 330,000 ha in Mediterranean countries are in protected cultivation, of which 190,000 ha are greenhouse production (Jouet, 2001). The leading countries in the region are Spain, Turkey, and Italy, followed by France, Israel, and Greece. Greenhouse culture also has constraints, such as pollination. Pollination (the deposition of an adequate amount of viable pollen on the receptive stigma at the right time for fertilization) is essential to most of the horticulture plants grown in greenhouses to achieve a reasonable fruit weight and quality. A great portion of the fruit quality in the extra-early horticulture (out-of-season fruits and vegetables) depends on efficient pollination.

From a strictly economic point of view, the importance of pollination of greenhouse crops may be seen in table 3.1. Approximately 45% of greenhouse crop production value is attributed to pollination. These data are from the 2002 Official Report of the Andalusian Agriculture and Fisheries Council in Almería (southeast of Spain), where more than 20,000 ha are cultivated under plastic greenhouses.

Table 3.1 Pollinator dependency factors for certain crops. Calculated according to Southwick & Southwick (1992).[1]

Plant	Production (metric tons)	Production Value (Euros × 1000)	Pollinator Dependency Factor	Generated Value per Pollinator (Euros × 1000)
Watermelon	200,210	54,057	0.4	21,622.8
Melon	163,024	84,772	0.6	50,863.2
Zucchini	229,352	144,492	0.6	86,695.2
Cucumber	262,200	167,808	0.1	16,780.8
Eggplant	70,200	35,100	0.6	21,060
Tomato	806,736	572,783	0.6	343,669.8
Pepper	542,925	352,901	0.2	70,580.2
Green bean	64,970	70,817	0.01	708.17
Total	2,339,617	1,482,730		611,980.17

[1] These authors calculated the dependency factor of each crop by assuming that the pollinators do not exist.

From a taxonomic point of view, there are four families of plants that are grown in greenhouses for commercial farming: Solanaceae (tomatoes, peppers, and eggplants), Cucurbitaceae (melons, watermelons, zucchini, and cucumbers), Rosaceae (strawberries), and Leguminosae (green beans). This list varies from one country to another, and some of these crops are not grown in all places. For example, greenhouse strawberries are not commercially important in Spain, but they are in Israel and Turkey. All four plant families have their own peculiar floral biologies, which are important to consider when determining the need for pollination and which pollinator to use.

Influence of Floral Biology of Greenhouse Plants on Pollinators

Nectar Content

Honey bees and bumble bees (the most frequent pollinators used to pollinate in greenhouses) are attracted to pollen and nectar and especially to nectar sugars. Several studies have emphasized the importance of nectar volume, whereas others underlined the importance of the concentrations or relative amounts of specific sugars (Baker & Baker, 1983; Kevan, 2003). Nectar is a mostly aqueous combination of a number of substances (Baker & Baker, 1983), mainly sucrose, glucose, and fructose. Other carbohydrates—including arabinose, galactose, mannose, gentiobiose, lactose, maltose, melibiose, trehalose, melezitose, raffinose, and stachyose—have also been identified in the nectars

Table 3.2 Main sugars found in nectars of the extra-early horticulture crops (Guerra-Sanz et al. 2005). All analyses were carried out with at least two varieties per crop species, using HPLC[1] except for the reference for strawberry.

Species	Sucrose	Glucose + Fructose	Other Detected Sugars	Nectar Type
Pepper	Yes	Yes	Raffinose traces	Glucose-fructose prevalent (sucrose/ hexoses ratio depends on the variety and anthesis time)
Melon	Yes	Yes		Sucrose prevalent (sucrose/ hexoses ratio depends on flower gender)
Watermelon	Yes	Yes	Raffinose, Stachyose	Sucrose prevalent (sucrose/ hexoses ratio depends on flower gender)
Zucchini	Yes	Yes	Raffinose, Stachyose traces	Sucrose prevalent (sucrose/ hexoses ratio depends on the variety and flower gender)
Cucumber	Yes	Yes		Sucrose prevalent (sucrose/ hexoses ratio depends on the variety and flower gender)
Strawberry	Yes	Yes		Glucose and fructose prevalent (Grünfeld et al., 1989)

[1]SugarPack 1 column, mobile phase water, 0.5 ml/min, 90°C of column oven temperature, detection by Index of Refraction.

of some flowers (Kevan, 2003). The various types of nectars can be ordered into three groups according to sugar content: sucrose prevalent, glucose and fructose prevalent, and almost equal amounts of sucrose, glucose, and fructose (table 3.2). Many other chemical substances have been detected in different nectars (Kevan, 2003), such as amino acids, enzymes, mineral ions, and so forth.

Other Flower Attractions for Pollinators

Such flower attributes as size, color, flower organs, nectar guides on the petals, nectar volume, nectar composition, and amount of pollen are considered to be important factors in attracting honey bees and bumble bees, and as such they can affect visitation frequency

(Dobson et al., 1990; Fahn, 1979; McGregor, 1976). For instance, corolla color reflectance of zucchini shows dimorphism between flower genders (J. M. Guerra-Sanz, A. Roldan, & A. Mena, personal communication, 2006), which may contribute to the selective foraging behavior found in bumble bees. Furthermore, the detection of the corolla's nectar guides by pollinators might be influenced by the plastic cover of the greenhouse, because some materials reduce the ultraviolet (UV) of the daylight spectrum, as explained later.

Recent studies indicate that the chemical components contributing to a flower's fragrance also play an important role in the attractiveness of flowers to bees (Henning et al., 1990; Masson et al., 1993; Matile & Altenburger, 1988; Pham-Delegue et al., 1989). The chemical composition of flower volatiles may also affect bee behavior. Olfactory signals are rapidly learned, indicating that foraging behavior results from the association of plant allelochemicals acting as chemosensory cues for the bees (Pham-Delegue et al., 1990). Moreover, the fact that in some cases bees are more attracted to flowers with a meager level of nectar than to those with high levels indicates that the olfactory signal(s) may be a dominant factor controlling bee behavior. Therefore, bee behavior is controlled by the integration of both perceived cues, such as color and/or fragrance, and the actual amount of a reward, such as pollen and nectar. Current studies have focused on describing the importance of each factor and the interaction between them in the attractiveness of flowers to bees. The effects of flower rewards and cues have been investigated by our research group (Mena Granero et al., 2004; 2005a, 2005b), but it would be premature to speak about all horticulture-cultivated species.

Impact of Physical Properties of Greenhouses on Pollinators

Greenhouse Covers

Rather than review all types of greenhouses, I summarize here the main ecological consequences that greenhouses impose on pollinators and their ability to pollinate. Generally, greenhouses are covered with glass or different types of plastic films, such as polyvinylchloride (PVC), polycarbonate (PC), and polyethylene (PE). PE is the most popular of the plastic materials. Besides other aspects, all materials differ in their transmission of UV light (wave length between 300 and 400 nm). UV-blocking plastic films help in decreasing the population levels of harmful insects in the crops (Costa et al., 2002). However, UV light is an important component of bee vision and orientation (Peitsch et al., 1992), and the degree of UV transmission through the greenhouse covering affects the behavior of bees used as pollinators. Under glass with a good UV transmission (up to 80%), bees behave normally. Under PVC or PC, with a very low UV transmission (less than 3%), they perform poorly, at least until they "learn" to cope with the lack of UV vision (see later in the chapter).

A new integrated pest management (IPM) strategy in greenhouses is to block UV radiation for pest control (Soler et al., 2006; figure 3.1). In tomato greenhouses,

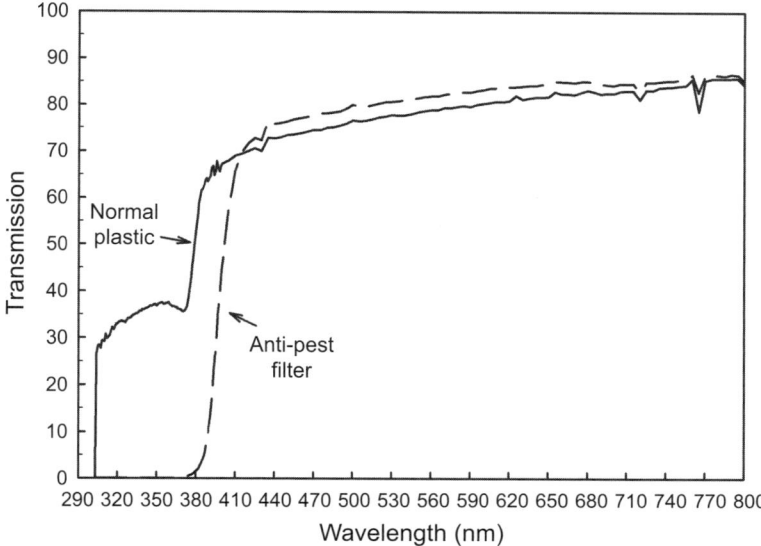

Figure 3.1 Percent of light transmission of two greenhouse plastics: "normal plastic" without a UV filter (black line) and "antipest" plastic with a UV filter (black curve).

however, reduced radiation interfered with the bees' navigation ability and thus reduced their activity (Dag & Eisikowitch, 2005). In a bumble bee experiment, different behaviors were noticed depending on the type of plastic cover used (Soler et al., 2006). Under UV-absorbing plastic significantly more bumble bees appeared at the nest entrance without flying to forage than was the case under plastic without UV filters. The bumble bees that did fly under UV-absorbing plastic spent more time at the platform nest entrance before flying. Bumble bees that returned from a flight spent more time at the entrance before actually reentering the nest. Under UV-blocking material, daily activity started slightly later in the day than under material without UV filters.

Morandin and co-workers (2001) compared four types of PE, one with a high degree of UV transmission (called CT), the other three types transmitting only a small fraction (down to 0%) of the UV light. They found that the bees under the CT plastic made twice as many foraging trips as bees under the other three types. Moreover, fewer bees got lost under the CT plastic (136% more bees remaining after 10 days). Similarly, bumble bee colonies performed better, as measured by foraging visits to flowers, under UV-blocking plastics when some normal (UV-containing) daylight entered the greenhouse, for instance, through pieces of gauze screen.

Dyer and Chittka (2004), using artificial tomato flowers offered at a distance of about 1 meter under UV+ and UV− conditions, showed that bumble bees indeed detected the presence or absence of UV light. However, the bees were able to find the flowers under both conditions, presumably after learning to recognize flowers in the absence of UV light.

Enrichment of the Atmosphere With Carbon Dioxide

In modern greenhouses, the CO_2 level is artificially increased to stimulate the growth of plants (up to three times the natural level of about 360 ppm). In some cases, measurements of CO_2 close to the gas outlets can be as high as 10,000 ppm. The activity and development of bumble bee colonies placed close to the outlets are negatively affected at values above 1,000 ppm. Research has shown that from 1,000 ppm upward bees become less active, and at around 5,000 ppm the first larval and adult mortality occur (van Doorn, 2006). Colony mortality occurs above 15,000 ppm. Therefore, nest boxes must be placed away from the gas outlets (not underneath them and at least 1 m away from them), or the outlets near the nest boxes must be closed. Another interesting effect of CO_2-enriched atmosphere was noticed in melons, in which CO_2 levels were found to affect the floral rewards by increasing nectar sugar concentration and possibly honey bee activity (Dag & Eisikowitch, 2000).

Greenhouse Temperature and Humidity

The amount of incoming pollen is influenced not only by the availability of pollen inside the greenhouse but also by bee foraging activity. Bumble bee workers usually do not forage at temperatures below 10°C (Heinrich, 1979). However, compared with other bees, including honey bees, they forage at relatively low temperatures. For this reason, they are highly esteemed as pollinators of protected crops growing under adverse climatic conditions (see, e.g., Abak et al., 1997, and Ercan & Onus, 2003, for pepper; Dasgan et al., 2004, for tomato; Abak & Dasgan, 2005, for eggplant). The foraging activity of bumble bees is also affected by high temperatures (above 30°C). In plastic greenhouses, maximum temperatures often rise above that level (up to around 40°C; Abak & Dasgan, 2005). Bumble bees limit foraging when the temperature rises above 32°C (Kwon & Saeed, 2003, for *Bombus terrestris*), although we have observed bumble bee workers foraging at 45°C (J. M. Guerra-Sanz & A. Roldán-Serrano, personal communication, July 2004). They are able to fly at air temperatures up to 35°C, but they prefer to stay at the nest to ventilate the brood. Above 32°C, bumble bee workers not only stop foraging and start ventilating the brood, but they also stop feeding the larvae (Heinrich, 1979; Vogt, 1986). Bumble bee larvae can starve for a considerable length of time (up to 2–3 days) before they die; however, a period of starvation results in a more prolonged developmental time (Plowright & Pendrel, 1977; Sladen, 1989; Sutcliffe & Plowright, 1990). Nevertheless, because hot days usually are accompanied by periods with moderate temperatures during early morning and late afternoon (Abak & Dasgan, 2005), usually the pollination activity and, therefore, the pollen intake of the colonies will not be completely blocked. At around 40°C bees prevent their own bodies from overheating by becoming inactive, and they stop fanning. As long as they are ventilating the nest, they are able to keep the brood temperature equal to, or just above (1–2°), ambient temperature (Heinrich, 1979; Vogt, 1986), but at temperatures over 40°C they are not able to cool the brood below ambient temperature. Vogt (1986) suggested that the reason is that little or no evaporative cooling is used. It is well known that some other social insects, such as honey bees and wasps, cool the nest by evaporating water that has been collected for that purpose (e.g., Wilson, 1971). Although there is no clear agreement on the temperature threshold limit, it has been indicated that an ambient temperature of 40°C is about the maximum temperature at which bumble bee colonies can survive, on condition that a sufficient energy supply is available (van Doorn, 2006).

Pollen production occurs without any problems in regularly heated greenhouses in cold winter regions, such as in Holland. However, the amount and quality of pollen decreases in regions with a mild winter climate (Abak et al., 1997). For example, in Turkey or Spain, heating is used to prevent frost only at particular times. Consequently, there are substantial fluctuations in greenhouse temperatures (at night in winter temperatures are low and in the daytime in spring temperatures are high inside the greenhouse; Abak et al., 1995). Other important problems from a climatic point of view are high humidity due to inadequate ventilation and low light permeability due to low quality of plastic covers. Both circumstances are limiting for quality pollen production. Because of the problems mentioned, greenhouse pollination has gained in relevance for research in the Mediterranean greenhouse region.

A final consideration about temperature and bee behavior has been recently raised by Dyer et al. (2006), who demonstrated that bees prefer warm nectar to cool nectar and that they are capable of using color to predict floral temperature before landing. Floral color signals are used by pollinators as predictors of nutritional rewards, such as nectar. But as insect pollinators often need to invest energy to maintain their body temperature above the ambient temperature, floral heat might also be perceived as a reward. Bumble bees (*Bombus terrestris*) prefer to visit warmer flowers and can learn to use color to predict floral temperature before landing. In what could be a widespread floral adaptation, plants may modulate their temperature to encourage pollinators to visit.

Bees Used for Pollination in Greenhouses

Bumble Bees: Commercial Use and Global Distribution

A controversy has recently appeared over the use of certain bees in greenhouses (Velthuis & van Doorn 2006): Is the domestication of a wild animal (or plant) enough reason to distribute it all around the world? Or should a species remain in only its original ecosystem or at least within its normal range limits? If so, who will make that decision? And how would such a decision be enforced worldwide? The use of nonnative bumble bees illustrates this controversy (see also chapter 9 in this volume). On the one hand, there are opinions from conservationists; on the other hand, there are requests from tomato farmers and companies that commercially rear bumble bees. These latter individuals argue that they should be allowed to freely trade in bumble bees, with no concern about introducing an exotic species. Tomato farmers know that bumble bees pollinate tomatoes cheaper and better than any other technique.

It would be naïve not to recognize that unbridled exportation of a certain insect species to a new place certainly has environmental risks, despite whether colonization of this new insect species had been assessed previously as causing no harm to natural competitors (Nagamitsu et al., 2007). For example, European bumble bee diseases (*Nosema bombi, Locustacarus* spp., etc.) are claimed to have been spread to native bee species in Japan and New Zealand (Thorp, 2003) on account of the importation of *B. terrestris* nests. Or the diseases may have been spread by rearing species native to Europe (e.g., *B. ignitus*) in the same commercial facilities as those used for *B. terrestris* and then sending those commercially reared native bees back to Japan, only now carrying the common diseases of *B. terrestris*.

Bombus terrestris: A "Model" Pollinator for Greenhouse Use

Bumble bees have an advantage over honey bees, especially in their foraging ability, because bumble bees can perform "buzz" pollination (Buchmann, 1983; Corbet et al., 1988; Harder & Barclay, 1994). Because of their special attributes, bumble bees are often used as greenhouse crop pollinators. It has been estimated that the bumble bees sold in 2004 were around 930,000 colonies of the Eurasian *B. terrestris*, around 55,000 colonies of the North American *B. impatiens*, and a few thousand colonies of the Eurasian *B. lucorum*, East Asian *B. ignitus*, and North American *B. occidentalis* (Velthuis & van Doorn, 2006). Therefore, from a strictly economic standpoint, bumble bee rearing and marketing is a "big" business, worth several billion dollars. The production of bumble bee colonies has been incorporated into the larger agribusiness chain, but not without problems—including the exportation of new species to a continent or island, as previously mentioned.

It has been claimed that a bumble bee hive living in a greenhouse is subjected to the same restrictions as in a laboratory or rearing facility (Velthuis & van Doorn, 2006), although some differences have been also raised. For example, the number of workers appears to diminish more quickly in the greenhouse than in a rearing facility for two main reasons: the poorer quality of food in a greenhouse compared with a rearing facility and more acute temperature fluctuations in the greenhouse.

There is great variation with respect to the maximum number of workers, queens, and males in colonies of each bumble bee species. Typically, colonies used for commercial pollination purposes are selected when the worker population reaches 50 individuals (Velthuis & van Doorn, 2006), although these worker populations are small compared with natural colony sizes. Colonies of *B. terrestris* may produce up to 400 workers, with up to 200 queens and several hundred males. These figures are comparable to *B. occidentalis*. Colonies of *B. impatiens* become almost two times larger, and those of *B. ignitus*, and certainly of *B. lucorum*, remain around one-half the size of *B. terrestris* colonies (Velthuis & van Doorn, 2006).

In *B. terrestris*, as in any bumble bee species, only queens hibernate. Single queens found a colony after hibernation. Initially, eggs are laid by the queen, and the subsequent emergence of adults occurs in a stepwise fashion, such that a first, second, and third period of egg laying can be distinguished. During the first period of egg laying, the queen lays fertilized (female-producing) eggs and behaviorally dominates the workers emerging from these eggs. These early emerging workers assist the queen in brood care. At some point later in the colony life cycle, the queen switches to laying unfertilized (male-producing) eggs ("switch point"), and, after that period, she loses control over worker reproduction ("competition point"). Usually, these events happen during the third period of egg laying, but they may also appear earlier. From the competition point onward, the most dominant workers lay unfertilized eggs, eat eggs laid by the queen or fellow workers, and may also attack each other or the queen. Moreover, during this competition phase, queens may be reared from the last fertilized eggs (van Honk & Hogeweg, 1981; van der Blom, 1986; van Doorn & Heringa, 1986; Duchateau & Velthuis, 1988; Röseler & van Honk, 1990). Colonies of *B. impatiens, B. occidentalis, B. lucorum*, and *B. ignitus* develop in a similar way (Hannan et al., 1997; Asada & Ono, 2000), yet in *B. impatiens* the level of aggression during the competition phase has been found to be lower than in *B. terrestris* (Pomeroy, 1981; Cnaani et al., 2002).

In the greenhouse, the colonies of the various bumble bee species develop as might be expected on the basis of the aforementioned laboratory experience. There is, however, a difference in the life expectancy of the workers: workers from free-foraging colonies have much shorter average life spans than workers from colonies kept in the laboratory. This is mainly due to the loss of foragers (Brian, 1952; Garófalo, 1978; van Doorn & Heringa, 1986; Küpper & Schwammberger, 1994; Katayama, 1996). The greater loss of foragers in the field can be explained in part by faster morphological and physiological aging (e.g., through wing wear; Cartar, 1992). This also applies to bumble bee foragers in a greenhouse environment. Furthermore, when nest boxes have been recently placed in a greenhouse and the vents are left open, foragers may get lost during their first orientation flight (if they leave the greenhouse through the vents and are unable to find their way back). Foragers may also leave the greenhouse in an attempt to collect food from outside the building (see later in the chapter). Furthermore, such foragers might get lost when the vents have been closed before they return. Other factors that influence the life expectancy of individual bees or the colony as a whole include food quality and availability, environmental conditions, the position of the hives, the occurrence of predators and parasites, and the use of pesticides.

Honey Bees: Management and Use of Hives in a Greenhouse

Honey bees have long been recognized as excellent pollinators of many plants, including many crops such as peach, almond, pear, apple, and melon (McGregor, 1976; Corbet et al., 1991). The main features that place honey bees ahead of other insects and other bees as pollinators are flower constancy, colony size, and recruitment behavior. When an individual worker bee forages, it normally visits only one type of flower at a time and ignores other flower types, even if they are more rewarding or closer. For example, a honey bee in an orchard might visit only apple tree flowers and ignore dandelion flowers, whereas another honey bee from the same hive might visit only dandelion flowers and ignore apple tree flowers. Each bee stays constant to one flower species, although it may change to another if the flowers it is visiting become less rewarding. This behavior is pronounced in honey bees and bumble bees and sometimes even in butterflies (Free, 1963, 1970; Goulson, Stout, & Hawson, 1997). Bees have difficulty collecting pollen from plant species if the pollen grains do not fit together well (Vaissière & Vinson, 1994), so it is possible that pollen collection from more than one species would be difficult because grains might not assemble easily. However, this pollen-collecting problem does not explain why flower constancy is also shown by bees collecting only nectar.

Although honey bees are excellent pollinators of many plants, they are not always the best pollinators for all plants. They have short tongues and cannot reach the nectar in deep flowers. For such flowers as those of fava beans (Kirk, 2004), long-tongued bumble bees are often better pollinators. Honey bees also require higher temperatures (> 12°C) in order to forage, so bumble bees may work better or forage longer in cool weather (Willmer, 1983).

From a commercial point of view, honey bee hives are introduced into greenhouses mainly for pollination of "short"-blooming crops, such as melon and watermelon. However, for long-blooming crops such as pepper, cucumber, or zucchini, there may be problems with the maintenance of a hive because of the stress imposed by the greenhouse temperature fluctuations.

In addition, some of the physical constraints imposed by the enclosures have to be taken into account before introducing honey bees into a greenhouse. For example, airflow direction and level with respect to the hive location affect bee pollination activity and fruit set in melon. This phenomenon can be explained by the tendency of bees to fly upwind (Dag & Eisikowitch, 1995).

Crop Pollination

Floral Biology of the Most Common Horticultural Crop Species in Greenhouses

Some crops set fruit parthenocarpically (e.g., some cucumbers and zucchini varieties), but this process does not always occur in the particular variety that is doing well in the market. Fruit and seed set are especially dependent on successful pollination when the species under consideration cannot or must not be automatically self-pollinated. This is true if flowers have any adaptation for the avoidance of selfing: spatial (intrafloral herkogamy, monoecious, or dioecious dicliny) or temporal (dichogamy: protandry, protogyny) separation of stigma and pollen presentation or self-incompatibility (table 3.3).

For crop production, some inconveniences can arise, especially in the cases of plants that exhibit andromonoecy, for example, eggplant and melon. In these cases, the abundance of pollen might be a hindrance for crop production, due to the selective foraging behavior that male flowers (which usually are more abundant than female or hermaphrodite flowers in these species) can exert on pollinators. The permanence of this trait

Table 3.3 Floral biology and some important traits for pollination of the extra-early horticultural species.

Species	Floral Biology	Nectar	Pollen	Special Traits
Tomato (*Lycopersicon esculentum* Mill.)	Inflorescence Hermaphrodite flowers (Truss)	Absent	Poricide anther	
Pepper (*Capsicum annuum* L.)	Hermaphrodite (a certain degree of dichogamy)	Present	Anther dehiscent	Nectar variability according to cultivar and season
Eggplant (*Solanum melongena* L.)	Andromonoecy	Absent	Poricide anther	
Melon (*Cucumis melo* L.)	Andromonoecy or Monoecy	Present	Anther dehiscent	
Watermelon (*Citrullus lanatus*)	Monoecy	Present	Anther dehiscent	
Zucchini (*Cucurbita pepo* L.)	Monoecy	Present	Anther dehiscent	Floral anthesis very short (9 hours)
Cucumber (*Cucumis sativus* L.)	Monoecy	Present	Anther dehiscent	Cultivars with parthenocarpy
Strawberry (*Fragaria x ananassa*)	Inflorescence Hermaphrodite flowers	Present	Anther dehiscent	

(andromonoecy) is usually sought by plant breeders because of its convenience for making hybrids, and breeders would be reluctant to change this trait. Other characteristics that would facilitate pollination are the amelioration of nectar production and production of scent volatiles, although these two traits are not usually targeted by breeders for improvement of any cultivar.

Tomato Lycopersicon sculentum *(Mill) (Family Solanaceae)*

A revision on tomato pollination has recently been reported (Westerkamp & Gottsberger, 2000), which highlights the need for vibration to release the pollen from the tomato anthers. Anthers form a cone in which each anther is interconnected only by hairs. Pollen is shed into this cone and may escape only through a common opening at the tip, which is partly occluded by the style. To release pollen, the cone has to be vibrated by bees capable of performing "buzz" pollination. This pollination by sonication is required even for self-pollination.

Tomato flowers do not produce nectar, although it is still uncertain whether this is a trait found in wild *Lycopersicon* or whether it is due to domestication (Rick, 1950). Therefore, attractiveness to tomato flowers is mainly provided by pollen odor (Dobson & Bergström, 2000) and visual cues. The pollen odor originates from the pollenkitt, which contains volatiles that belong to the same chemical classes that are found in flower scent and that occur in species-specific mixtures. However, pollen odors differ from odors of other floral parts (Dobson & Bergström, 2000). Pollen-feeding insects can perceive pollen odor and use it to discriminate between different pollen types and host plants.

Pepper (Capsicum annuum L.) *(Family Solanaceae)*

Compared with tomato pollination, very little research has been done on pepper pollination, but it has been the subject of several controversial studies. For example, McGregor (1976) stated that peppers and other members of the Solanaceae are noted for their low attractiveness to honey bees. On the other hand, many researchers suggested that honey bees, thrips, and ants play a role in the cross-pollination of these flowers (Rabinowitch et al., 1993). In fact, the actual measurements of outcrossing under field conditions range from 2% to over 90%, depending on locality, environment, and space between plants (Pickersgill, 1997).

Nectar-sugar composition in pepper flowers is contradictory because old studies (Martin et al., 1932) indicated only the presence of glucose, although presence of sucrose was unexplored. Rabinowitch et al. (1993) showed that pepper nectar contains only fructose and glucose, whereas Roldán Serrano and Guerra-Sanz (2004), using a different analysis technique, found fructose, glucose, and sucrose. However, different pepper varieties have been analyzed in each case, and therefore a clear result would need further analysis. Additionally, great discrepancies have been found between Rabinowitch et al. (1993) and our results (Roldán Serrano & Guerra-Sanz, 2004) on nectar volume per flower. Again, these significant differences may be due to different pepper varieties used in each study.

Bee visits to pepper flowers indicate that they forage for nectar, trying to reach the bottom of the corolla where nectar drops are located (Rabinowitch et al., 1993). Because of flower morphology, when the bees extend their tongue to reach the bottom of the flower, at the same time their bodies touch the anthers, provoking the release of the pollen from mature anthers. Pollen can be deposited by a bee on the stigma in a greater amount than in self-pollination due to the spatial disposition of sexual organs in pepper flowers. Fruits obtained from bee-visited flowers had a consistently greater number of seeds than self-pollinated flowers, and they were also greater in size, enhancing their market quality (Roldán Serrano & Guerra Sanz, 2006).

Several pollinators have been used for pepper pollination in both commercial and research trials: (1) honey bees (de Ruijter, van den Eijnde, & van der Steen, 1991; Kubisová & Háslbachová, 1991; Dag & Kammer, 2001), (2) bumble bees, *Bombus terrestris* (L.) and

B. impatiens (Cr.; Abak et al., 1997; Meisels & Chiasson, 1997; Shipp et al., 1994; Dag & Kammer, 2001), (3) the solitary bee, *Osmia cornifrons* (Radoszkowski et al., 1991), and (4) the fly, *Eristalis tenax* (Jarlan et al., 1997).

From all of these studies on pepper, it is clear that the activity of any pollinator improves the quality and/or quantity of the fruit obtained compared with controls of self-pollination. Abak et al. (1997), for example, stated that the average yield, weight of fruit, diameter of fruit, and the number of seeds increased by 4.0%, 10.0%, 6.0%, and 12.5%, respectively, in the bumble bee-pollinated group of peppers compared with the control group in greenhouse trials. Similarly, in commercial greenhouses the average early and total yields of peppers increased by 29.6% and 22.4%, respectively, when pollinated by insects; fruit weight, diameter, volume, and flesh thickness were also positively influenced. Meisels and Chiasson (1997) reported that a colony of 30–40 *B. impatiens* was evaluated for its effectiveness as a sweet pepper pollinator (*Capsicum annuum* L. var. *grossum* cv. Superset) inside a screened greenhouse. *Bombus impatiens* activity in the nest and on flowers, individual *B. impatiens* flights, and the number of seeds produced per fruit were recorded during two periods: June 30–July 14 and August 4–18, 1995. Though activity continued inside the nest, *B. impatiens* worker foraging and colony size both decreased from one period to the next, yet there was no significant difference between periods for the number of seeds produced per fruit. Growers can expect effective pollination of sweet peppers by using *B. impatiens* throughout the growing season even with as few as 3 foragers per 425 plants (i.e., 176 *B. impatiens* foragers per hectare).

The fruit setting of sweet pepper is readily achieved in the summer without any pollinator in the greenhouse, but commercially viable fruit yields can be difficult to attain in the spring and autumn. The question has been raised whether insect pollination would improve fruit set in these cooler seasons. In autumn 1989 and spring 1990, experiments were carried out to compare fruit set of sweet pepper in greenhouses pollinated by honey bees with fruit set in greenhouses without bees. In both experiments, larger and heavier fruits with more seeds and fewer malformed fruits appeared after honey bees pollinated in the greenhouses. Nowadays, the use of honey bees to pollinate sweet pepper is a common practice in the Netherlands (de Ruijter et al., 1991).

*Watermelon (*Citrullus lanatus *[(Thunb.) Matsum. and Nakai]) (Family Cucurbitaceae)*

Watermelon has staminate (male) and pistillate (female) flowers in the same plant. Sedgley and Buttrose (1978) reported that the proportion of staminate flowers in watermelon increased together with temperature and that light did not greatly influence floral sex expression. On the other hand, Hawker and colleagues (1983) asserted that flower anthesis in watermelon lasts 2 days at 25°C, whereas at 30°C, they are withered by the end of the first day. However, stigma exudates containing fructose, glucose, sucrose, and polysaccharides were more noticeable at 30°C and also increased with age, allowing more attraction of bees. Increased pollination was also a result of an increase in secretion of compounds by the stigma.

Studies of watermelon pollination have been increasing recently because of the increasing interest in "seedless" varieties. The pollination and fruit set of traditional diploid ($n = 22$) watermelon varieties have been well studied. However, pollination and fruit management of the crop has become complex due to the introduction of commercial triploid ($n = 33$) varieties ("seedless" watermelon). Triploid watermelon needs to be pollinated and "fertilized" with viable pollen, yet triploid pollen is not viable. Therefore, some diploid plants must be grown together with the triploids in order to have enough viable pollen to obtain a reasonable yield. Real fertilization between diploid pollen and triploid ovules is not accomplished because there is no compatibility of chromosome number in the split of chromosome pairs. As a result, embryos are aborted (remaining in the fruit as empty seeds), but fruit is set, and growth and ripening starts as soon as enough diploid pollen is deposited in the stigmas. As a consequence of all this, a high pollination activity is needed for a triploid watermelon culture. Also, it is important to realize the competition established between diploid and triploid flowers in the same spot, because triploid flowers are bigger than diploid ones, having more nectar and pollen than the diploid, even though triploid pollen is not viable for fertilization (although the bees do not know that). One strategy aimed at "boosting" triploid watermelon pollination is the addition of honey bee brood pheromone inside the hives (Pankwit, 2004), and this technique has been already applied with great success (Guerra-Sanz & Roldán Serrano, 2007), resulting in a better fruit size and sugar concentration.

Watermelon flower attractiveness to pollinators has been studied (Wolf et al., 1999), including not only commercial diploid varieties but also some hybrids between cultivated and wild *Citrullus* spp. Differences are found among varieties, including daily nectar volume variation, sugar composition, and the number of honey bee visits to each variety. The frequency of bee visitations depends on watermelon genetics, as well as on environmental factors, such as temperature. Bee visits to watermelon flowers usually peak in the morning hours, because high temperatures later in the day negatively influence bee activity. Another factor affecting the number of bee visits is competition between the various species in a given area. Wolf et al. (1999) stated that genetic variability in attractiveness to honey bees was found within *Citrullus*, although floral attributes analysis indicated no genetic variability in flower size, amount of pollen grain, or nectar volume. However, differences were observed in the concentration of sucrose and total sugars in the nectar. A positive relationship was found between attractiveness to bees and nectar sugar concentration, suggesting that this characteristic is one of the parameters responsible for variability in attractiveness to bees.

Zucchini: Cucurbita pepo *(L.) (Family Cucurbitaceae)*

Zucchini is grown all over the world. Because cultivation in the field is restricted to spring and summer, the out-of-season production has to be maintained under protected cultivation, as in greenhouses.

Zucchini is a monoecious plant, having male and female flowers produced on the same plant. Its floral biology has been studied extensively (Nepi & Paccini, 1993; Roldán

Serrano & Guerra-Sanz, 2005). Honey bees and bumble bees have been used to pollinate zucchini in greenhouses (Guerra-Sanz et al., 2005), and pollination results in good fruit quality compared with fruit produced via parthenocarpic fertilization induced by growth regulators (Roldán Serrano et al., 2002; Guerra-Sanz et al., 2004).

Seed Crops

Crops cultured specifically for seed production are seldom grown in greenhouses. For the few that are grown in greenhouses, normally seed companies prefer manual pollination. Moreover, breeders try to avoid allowing any pollinator insect into the greenhouse facility during the pollination period to avoid uncontrolled pollinations. Only in a few cases can pollination by insects be considered, for instance, to produce seeds of pure lines from a determined lineage. In this case, the plants can be pollinated by a few pollinators if the plants and the pollinators are "jailed" within a small greenhouse or netted area within a larger greenhouse. It has been demonstrated that bumble bee drones behave very well in these cases and that an entire bumble bee colony is unnecessary (van Doorn, 2006).

Interactions Between Pollination and Pest Management in Enclosed Systems

The use of bees, and especially bumble bees, for pollination has led to a decrease in the use of pesticides for crop protection (Velthuis & van Doorn, 2006). In general, an increased emphasis on IPM practices is more frequent among users of bumble bee hives because they can manage bumble bees to minimize interruptions of pollination due to pesticide applications.

However, there are some inconveniences for bees on account of the IPM practices, such as the catching of bees and bumble bees on sticky traps, which are commonly used for trapping pests inside a greenhouse. The two main sticky trap colors are yellow and blue (Gillespie & Vernon, 1990). Because of the color attraction to specific pests, the first is mainly used to trap whiteflies and the second to catch thrips. A great number of worker bumble bees (probably juvenile adults without experience) were caught in blue traps (personal observations). The same phenomenon also has been observed for honey bees with yellow traps (personal observations). Assessments of the reflection spectrum of these traps have not found any particular spectral peak that could be the cause of the attraction of the bees. We can only speculate that the sticky surface, made with a gum that reflects sunlight, produces crystal reflections that attract bees.

Conclusion

For my work group, the question of introducing an exotic species into a different ecosystem is personal. In fact, it is an ethical question, not a matter of ecology, biology, or even an economic issue. Of course, ecology, biology, or economy should also be considered in answering the question, though the solution is not on the scientific side but on the ethical side. Otherwise, the point at issue will remain without response forever, because people from both sides (conservationists vs. farmers plus rearing companies) will have arguments in favor of and against, which would extend our discussion *ad infinitum*. In chapter 9 of this book, the reader will find more on this matter from the point of view of biology and ecology, which could help to make up one's mind on the issue. Certainly, the answer to the problem is not an easy one, and probably the views of everyone involved must be obtained in order to find a good solution.

Increasing awareness for greenhouse pollination techniques and studies has occurred over the past years, many of them inspired by the introduction of new cultures in greenhouses, such as triploid watermelons. Our forecast for greenhouse cultivation is very positive due to the high incomes generated by extra-early horticulture crops. New and more thrilling events are yet to come, such as the search for new pollinators. Generally speaking, greenhouse culture is expanding all over the world, and, together with the use of pollinators, real advances in IPM practices are growing, which gives a very positive and hopeful scenario for this particular corner of agriculture.

Acknowledgments

Thanks to A. Roldan Serrano and A. Mena Granero for their help on pollination projects. Thanks to Agrobio, S. L., for its support for several pollination events and experiments. Some of the results showed here have been supported by the following national and regional project grants: INIA RTA03-087, INIA RTA2005-00046-00-00, PIA-03-032.

References

Abak, K., & Dasgan, H. Y. (2005). Efficiency of bumblebees as pollinators in unheated or anti-frost heated greenhouses. In J. M. Guerra Sanz, A. Roldán Serrano, & A. Mena Granero (Eds.), *First short course on pollination of horticulture plants* (19–29). Almería, Spain: CIFA La Mojonera.

Abak, K., Dasgan, H. Y., Ikiz, Ö., Uygun, N., Kaftanoglu, O., & Yeninar, H. (1997). Pollen production and quality in pepper grown in anti-frost heated greenhouses during winter and the effects of bumblebee (*Bombus terrestris*) pollination on fruit yield and quality. *Acta Horticulturae, 437*, 303–307.

Abak, K., Sari, N., Paksoy, M., Kaftanoglou, O., & Yeninar, H. (1995). Efficiency of bumble bees on the yield of eggplant and tomato grown in unheated greenhouses. *Acta Horticulturae, 412*, 268–274.

Asada, S., & Ono, M. (2000). Difference in colony development of two Japanese bumblebees, *Bombus hypocrita* and *Bombus ignitus* (Hymenoptera: Apidae). *Applied Entomology and Zoology, 35*, 597–603.

Baker, H. G., & Baker, I. (1983). A brief historical review of the chemistry of floral nectar. In C. E. Jones & R. J. Little (Eds.), *The biology of nectaries*. New York: Columbia University Press.

Brian, A. D. (1952). Division of labour and foraging in *Bombus agrorum* Fabricius. *Journal of Animal Ecology, 21,* 223–240.

Buchmann, S. L. (1983). Buzz pollination in angiosperms. In C. E. Jones & R. J. Little (Eds.), *Handbook of experimental pollination biology* (73–113). New York: Van Nostrand Rheinhold.,

Cartar, R. V. (1992). Morphological senescence and longevity: An experiment relating wing wear and life span in foraging wild bumble bees. *Journal of Animal Ecology, 61,* 225–231.

Cnaani, J., Schmid-Hempel, R., & Schmidt, J. O. (2002). Colony development, larval development and worker reproduction in *Bombus impatiens* Cresson. *Insectes Sociaux, 49,* 164–170.

Corbet, S. A., Chapman, H., & Saville, N. (1988). Vibratory pollen collection and flower form: Bumble-bees on *Actinidia, Symphytum, Borago* and *Polygonatum. Functional Ecology, 2,* 147–155.

Corbet, S. A., Williams, I. H., & Osborne, J. L. (1991). Bees and the pollination of crops and wild flowers in the European Community. *Bee World, 72,* 47–59.

Costa, H. S., Robb, K. L., & Wilen, C. A. (2002). Field trials measuring the effects of ultraviolet-absorbing greenhouse plastic films on insect populations. *Journal of Economic Entomology, 95,* 113–120.

Dag, A., & Eisikowitch, D. (1995). The influence of hive location on honeybee foraging activity and fruit set in melons grown in plastic greenhouses. *Apidologie, 26,* 511–519.

———. (2000). The effect of carbon dioxide enrichment on nectar production in melon under greenhouse conditions. *Journal of Apicultural Research, 39,* 88–89.

———. (2005). The effect of environmental conditions on bee pollination activity in greenhouses [Abstract, p. 65], Thirty-ninth Apimondia, International Apicultural Congress, Dublin, Ireland.

Dag, A., & Kammer, Y. (2001). Comparison between the effectiveness of honey bee (*Apis mellifera*) and bumble bee (*Bombus terrestris*) as pollinators of greenhouse sweet pepper (*Capsicum annuum*). *American Bee Journal, 141*(6), 447–448.

Dasgan, H. Y., Özdogan, A. O., Kaftanoglu, O., & Abak, K. (2004). Effectiveness of bumblebee pollination in anti-frost heated tomato greenhouses in the Mediterranean Basin. *Turkish Journal of Agriculture and Forestry, 28,* 73–82.

de Ruijter, A., van den Eijnde, J., & van der Steen, J. (1991). Pollination of sweet pepper (*Capsicum annuum* L.) in greenhouses by honeybees. *Acta Horticulturae, 288,* 270–274.

Dobson, H. E. M., & Bergström, G. (2000). The ecology and evolution of pollen odors. *Plant Systematics and Evolution, 222,* 63–87.

Dobson, H. E. M., Bergström, G., & Groth, I. (1990). Differences in fragrance chemistry between flower parts of *Rosa rugosa* Thumb (Rosaceae). *Israel Journal of Botany, 39,* 143–156.

Duchateau, M. J., & Velthuis, H. H. W. (1988). Development and reproductive strategies in *Bombus terrestris* colonies. *Behaviour, 107,* 186–207.

Dyer, A. G., & Chittka, L. (2004). Bumblebee search time without ultraviolet light. *Journal of Experimental Biology, 207,* 1683–1688.

Dyer, A. G., Whitney, H. M., Arnold, S. E. J., Glover, B. J. & Chittka, L. (2006). Bees associate warmth with floral colour. *Nature, 442,* 525.

Ercan, N., & Onus, A. N. (2003). The effects of bumblebees (*Bombus terrestris* L.) on fruit quality and yield of pepper (*Capsicum annuum* L.) grown in an unheated greenhouse. *Israel Journal of Plant Sciences, 51,* 275–283.

Fahn, A. (1979). *Secretory tissues in plants.* London: Academic Press.

Free, J. B. (1963). The flower constancy of honeybees. *Journal of Animal Ecology, 32,* 119–131.

———. (1970). The flower constancy of bumblebees. *Journal of Animal Ecology, 39,* 395–402.

Garófalo, C. A. (1978). Bionomics of *Bombus (Fervidobombus) morio*: 2. Body size & length of life of workers. *Journal of Apicultural Research, 17,* 130–136.

Gillespie, D. R., & Vernon, R. S. (1990). Trap catch of western flower thrips (Thysanoptera: Thripidae) as affected by color and height of sticky traps in mature greenhouse cucumber crops. *Journal of Economic Entomology, 83*(3), 971–975.

Goulson, D., Stout, J. C., & Hawson, S. A. (1997). Can flower constancy in nectaring butterflies be explained by Darwin's interference hypothesis? *Oecologia, 112,* 225–231.

Grünfeld, E., Vincent, C., & Bagnara, D. (1989). High-performance liquid chromatography analysis of nectar and pollen of strawberry flowers. *Journal of Agriculture and Food Chemistry, 37,* 290–294.

Guerra-Sanz, J. M., Roldán Serrano, A. (2007, April). Feromona larval de abejas (*Apis mellifera* L.) en la producción de sandía (*Citrullus lanatus* [(Thunb.) Matsum. et Nakai*)* triploide en invernadero [Honey bee' s (*Apis mellifera* L.) brood pheromone influence on triploid watermelon production (*Citrullus lanatus* [(Thunb.) Matsum. et Nakai*)* in greenhouse] [Abstract, p. 13]. *Resúmenes Congreso Nacional Sociedad Española de Ciencias Hortícolas,* XI Congreso Nacional de Ciencias Hortícolas, Albacete, Spain [Publication in Spanish: Actas de Horticultura 48: 214–217].

Guerra-Sanz, J. M., Roldán Serrano, A., & Mena Granero, A. (2004). Pollination of zucchini culture by bumblebees: Advance of results of quality production. In A. Lebeda & H. S. Paris (Eds.), *Cucurbitaceae 2004: Progress in cucurbit genetics and breeding research. Proceedings of the 8th Eucarpia Conference, July 12–17, 2004, Olomouc, The Czech Republic* (75–77). Olomouc, Czech Republic: Palacký University.

Guerra-Sanz, J. M., Roldán Serrano, A., Mena Granero, A. & Fernández López, C. (2005). Polinización de hortalizas extra-tempranas en el marco de los invernaderos de Almería. In J. M. Guerra-Sanz, A. Roldán Serrano, & A. Mena Granero (Eds.), *Primeras jornadas de polinización en plantas hortícolas* (103–120). Almería, Spain: CIFA La Mojonera.

Hannan, A., Maeta, Y., & Hoshikawa, K. (1997). Colony development of two species of Japanese bumblebees *Bombus (Bombus) ignitus* and *Bombus (Bombus) hypocrita* reared under artificial condition (Hymenoptera, Apidae). *Japanese Journal of Entomology, 65,* 343–354.

Harder, L. D. & Barclay, R. M. R. (1994). The functional significance of poricidal anthers and buzz pollination: Controlled pollen removal from Dodecatheon. *Functional Ecology, 8,* 509–517.

Hawker, J. S., Sedgley, R. M. R., & Loveys, B. R. (1983). Composition of stigmatic exudates, nectar and pistil of watermelon, *Citrullus lanatus* (Thunb.) Matsum and Nakai, before and after pollination. *Australian Journal of Plant Physiology, 10,* 257–264.

Heinrich, B. (1979). *Bumblebee economics.* Cambridge, MA: Harvard University Press.

Henning, J. A., Peng, Y. S., Montague, M. A., & Teuber, L. R. (1990). Honeybee (*Hymenoptera: Apidae*) behavioural response to primary alfalfa (*Rosales fabaceae*) floral volatiles. *Journal of Economic Entomology, 85,* 233–239.

Jarlan, A., de Oliveira, D., & Gingras, J. (1997). Pollination of sweet pepper (*Capsicum annuum* L.) in green-house by the syrphid fly *Eristalis tenax* (L.). *Acta Horticulturae, 437,* 335.

Jouet, J. P. (2001). Plastics in the world. *Plasticulture, 2*(120), 106–127.

Katayama, E. (1996). Survivorship curves and longevity for workers of *Bombus ardens* Smith and *Bombus diversus* Smith (Hymenoptera, Apidae). *Japanese Journal of Entomology, 64,* 111–121.

Kevan, P. G. (2003). The modern science of ambrosiology: In honour of Herbert and Irene Baker. *Plant Systematics and Evolution, 238,* 1–5.

Kirk, W. D. J. (2004). Plants for bees: Faba bean: *Vicia faba. Bee World, 85,* 60–62.

Kristjansson, K., & K. Rasmussen. (1991). Pollination of sweet pepper (*Capsicum annuum* L.) with the solitary bee *Osmia cornifrons* (Radoszkowski). *Acta Horticulturae, 288,* 173–179.

Kubisová, S., & Háslbachová, H. (1991). Pollination of male-sterile green pepper line (*Capsicum annuum* L.) by honeybees. *Acta Horticulturae, 288,* 364.

Küpper, G., & Schwammberger, K.–H. (1994). Volksentwicklung and Sammelverhalten bei *Bombus pratorum* (L.) (Hymenoptera, Apidae). *Zoologische Jahrbucher Systematik, 121,* 202–219.

Kwon, Y. J. & Saeed, S. (2003). Effect of temperature on the foraging activity of *Bombus terrestris* L. (Hymenoptera: Apidae) on greenhouse hot pepper (*Capsicum annuum* L.). *Applied Entomology and Zoology, 38,* 275–280.

Martin, J. A., Erwin, A. T., & Lounsberry, C. C. (1932). Nectaries of *Capsicum. Journal of Science, 6,* 277–285.

Masson, C., Pham-Delegue, M. H., Fonta, C., Gascuet, J., Arnold, G., Nicolas, G., et al. (1993). Recent advances in the concept of adaptation to natural odour signals in honeybee *Apis mellifera* L. *Apidologie, 24,* 169–194.

Matile, P. & Altenburger, R. (1988). Rhythms of fragrance emissions in flowers. *Planta, 174,* 242–247.

McGregor, S. E. (1976). *Insect pollination of crop plants* (USDA-ARS Agriculture Handbook No. 496). Washington, DC: U.S. Department of Agriculture.

Meisels, S., & Chiasson, H. (1997). Effectiveness of *Bombus impatiens* Cr. as pollinators of green-house sweet peppers (*Capsicum annuum* L.). *Acta Horticulturae, 437,* 425–429.

Mena Granero, A., Egea González, F. J., Garrido Frenich, A., Guerra Sanz, J. M., & Martínez Vidal, J. L. (2004). Single step determination of fragrances in *Cucurbita* flowers by coupling headspace solid-phase microextraction low-pressure gas chromatography–tandem mass spectrometry. *Journal of Chromatography-A, 1045,* 173–179.

Mena Granero, A., Egea González, F. J., Guerra Sanz, J. M., & Martínez Vidal, J. L. (2005a). Analysis of biogenic volatile organic compounds in zucchini flowers: Identification of scents sources. *Journal of Chemical Ecology, 31*(10), 2309–2322.

Mena Granero, A., Guerra-Sanz, J. M., & Egea González, F. J.(2005b). Química de la polinización [Pollination chemistry]. In J. M. Guerra-Sanz, A. Roldán Serrano, & A. Mena Granero (Eds.), *Primeras jornadas de polinización en plantas hortícolas* (31–48). Almería, Spain: CIFA La Mojonera.

Morandin, L. A., Laverty, T. M., Kevan, P. G., Khosla, S., & Shipp, L. (2001). Bumble bee (Hymenoptera: Apidae) activity and loss in commercial tomato greenhouses. *Canadian Entomologist, 133,* 883–893.

Nagamitsu, T., Kenta, T., Inari, N., Kato, E., & Hiura, T. (2007). Abundance, body size, and morphology of bumblebees in an area where an exotic species, *Bombus terrestris,* has colonized in Japan. *Ecological Research, 22,* 331–341.

Nepi, M., & Paccini, E. (1993). Pollination, pollen viability and pistil receptivity in *Cucurbita pepo*. *Annals of Botany, 72*, 527–536.

Pankwit, T. (2004). Brood pheromone regulates foraging activity of honey bees (Hymenoptera: Apidae). *Journal of Economical Entomology, 97*(3), 748–751.

Pardossi, A., Tognoni, F., & Incrocci, L. (2004). Mediterranean greenhouses technology. *Chronica Horticulturae, 44*(2), 28–34.

Peitsch, D., Fietz, A., Hertel, H., de Souza, J., Ventura, D. F., & Menzel, R. (1992). The spectral input systems of hymenopteran insects and their receptor-based colour vision. *Journal of Comparative Physiology A, 170*, 23–40.

Pham-Delegue, M. H., Etievant, P., Guichard, E., Marilleau, R., Duault, P., Chauffaille, J., et al. (1990). Chemicals involved in honey-bee-sunflower relationship. *Journal of Chemical Ecology, 16*, 3053–3065.

Pham-Delegue, M. H. P., Etievant, P., Guichard, E., & Masson, C. (1989). Sunflower volatiles involved in honeybee discrimination among genotypes and flowering stages. *Journal of Chemical Ecology, 15*, 329–343.

Pickersgill, B. (1997). Genetic resources and breeding of *Capsicum* spp. *Euphytica, 96*, 129–133.

Plowright, R. C. & Pendrel, B. A. (1977). Larval growth in bumble bees (Hymenoptera: Apidae). *Canadian Entomologist, 109*, 967–973.

Pomeroy, N. (1981). Reproductive dominance interactions and colony development in bumblebees (*Bombus* Latreille; Hymenoptera: Apidae). Unpublished doctoral dissertation, University of Toronto.

Rabinowitch, H. D., Fahn, A., Meir, T., & Lensky, Y. (1993). Flower and nectar attributes of pepper (*Capsicum annuum* L.) plants in relation to their attractiveness to honeybees (*Apis mellifera* L.). *Annals of Applied Biology, 123*, 221–232.

Rick, C. M. (1950). Pollination relations of *Lycopersicon esculentum* in native and foreign regions. *Evolution, 4*, 110–122.

Roldán Serrano, A., & Guerra-Sanz, J. M. (2004). Dynamics and sugar composition of sweet pepper (*Capsicum annuum* L.) nectar. *Journal of Horticultural Science and Biotechnology, 79*(5), 717–722.

Roldán Serrano, A., & Guerra-Sanz, J. M. (2005). Reward attractions of zucchini flowers (*Cucurbita pepo*, L.) to bumblebees (*Bombus terrestris* L.). *European Journal of Horticultural Science, 70*(1), 23–28.

Roldán Serrano, A., Guerra-Sanz, J. M., & Ortuño Izquierdo, M. J. (2002). Flower attractiveness to bumble-bees (*Bombus terrestris* L.) in zucchini (*Cucurbita pepo* L.). In D. N. Maynard (Ed.), *Cucurbitaceae, 2002* (343–348). Alexandria, VA: ASHS Press.

Röseler, P.-F., & van Honk, C. G. J. (1990). Castes and reproduction in bumblebees. In W. Engels (Ed.), *Social insects: An evolutionary approach to castes and reproduction* (147–166). Berlin, Germany: Springer.

Sedgley, M., & Buttrose, M. S. (1978). Some effects of light intensity, daylength and temperature on flowering and pollen tube growth in the watermelon (*Citrullus lanatus*). *Annals of Botany, 42*, 609–616.

Shipp, J. L., Whitfield, G. H., & Papadopoulos, A. P. (1994). Effectiveness of the bumble bee, *Bombus impatiens* Cr. (Hymenoptera: Apidae), as a pollinator of greenhouse sweet pepper. *Scientia Horticulturae, 57*, 29–39.

Sladen, F. W. L. (1989). *The humble-bee, its life-history and how to domesticate it.* Woonton, Hereford, UK: Logaston Press.

Soler, A., van der Blom, J., López, J. C., & Cabello, T. (2006). The effect of the absorbent UV plastic on the behaviour of *Bombus terrestris* in greenhouses; results of a bioassay. In J. M. Guerra-Sanz, A. Roldán Serrano, & A. Mena Granero (Eds.), *Second short course on pollination of horticulture plants* (258–261). Almería, Spain: CIFA La Mojonera.

Southwick, E. E. & Southwick, L., Jr. (1992). Estimating the economic value of honey bees (Hymenoptera: Apidae) as agricultural pollinators in the United States. *Journal of Economic Entomology, 85,* 621–633.

Sutcliffe, G. H., & Plowright, R. C. (1990). The effects of pollen availability on development time in the bumble bee *Bombus terricola* K. (Hymenoptera: Apidae). *Canadian Journal of Zoology, 68,* 1120–1123.

Thorp, R. W. (2003). Bumble bees (Hymenoptera: Apidae): Commercial use and environmental concerns. In K. Strickler & J. H. Cane (Eds.), *For nonnative crops, whence pollinators of the future?* (21–40). Lanham, MD: Entomological Society of America.

Vaissière, B. E. & Vinson, S. B. (1994). Pollen morphology and its effect on pollen collection by honey bees, *Apis mellifera* L. (Hymenoptera: Apidae), with special reference to upland cotton, *Gossypium hirsutum* L. (Malvaceae). *Grana, 33,* 128–138.

van Doorn, A. (2006). Factors influencing the performance of bumblebee colonies in the greenhouse. In J. M. Guerra-Sanz, A. Roldán Serrano, & A. Mena Granero (Eds.), *Jornadas de Polinización en plantas hortícolas* (2nd ed., 173–183). Almería, Spain: CIFA La Mojonera.

van Doorn, A., & Heringa, J. (1986). The ontogeny of a dominance hierarchy in colonies of the bumblebee *Bombus terrestris* (Hymenoptera, Apidae). *Insectes Sociaux, 33,* 3–25.

van der Blom, J. (1986). Reproductive dominance within colonies of *Bombus terrestris* (L.) *Behaviour, 97,* 37–49.

van Honk, C., & Hogeweg, P. (1981). The ontogeny of the social structure in a captive *Bombus terrestris* colony. *Behavioral Ecology and Sociobiology, 9,* 111–119.

Velthuis, H. H. W. & van Doorn, A. (2006). A century of advances in bumblebee domestication and the economic and environmental aspects of its commercialization for pollination. *Apidologie, 37*(4), 421–451.

Vogt, F. D. (1986). Thermoregulation in bumblebee colonies: 1. Thermoregulatory versus brood-maintenance behaviors during acute changes in ambient temperature. *Physiological Zoology, 59,* 55–59.

Westerkamp, C. & Gottsberger, G. (2000). Diversity pays in crop pollination. *Crop Science, 40,* 1209–1222.

Willmer, P. G. (1983). Thermal constraints on activity patterns in nectar-feeding insects. *Ecological Entomology, 8,* 455–469.

Wilson, E. O. (1971). *The insect societies.* Cambridge, MA: Belknap Press.

Wolf, S., Lensky, Y., & Paldi, N. (1999). Genetic variability in flower attractiveness to honeybees (*Apis mellifera* L.) within the genus *Citrullus. Journal of the American Society for Horticultural Science, 34,* 860–863.

4 Pollinating Bees Crucial to Farming Wildflower Seed for U.S. Habitat Restoration

James H. Cane

Background

Native plant communities and ecosystems of the western United States are increasingly besieged by invasive weeds and grasses of Eurasian origins. Federal land managers oversee 40 million ha in the Great Basin region of the western United States. Its valleys are cold desert steppe vegetated by grassy shrublands; these are interrupted by forested mountains. The Great Basin and Columbia Plateau ecoregions (figure 4.1) together include about one-half of North America's remaining 43 million ha of sagebrush (*Artemesia* spp.) plant communities (Wisdom et al., 2005). The health and extent of these communities are fast diminishing, with dire consequences for many plant and animal inhabitants. The most charismatic of these is the sage grouse, but 206 other species warrant conservation concern (Wisdom et al., 2005). Cheatgrass (smooth brome, *Bromus tectorum*), a flammable annual grass from the Mediterranean, has invaded 11 million ha of the Great Basin northward into the Columbia Plateau, an area equivalent to that of Cuba, Hungary, or Virginia. Cheatgrass is quite literally fueling a dramatic increase in the frequency and extent of wildfire across the region, degrading native plant communities and presumably their dependent native bee faunas. In Nevada alone, 1,000 wildfires burned 200,000 ha of rangeland from 2000 to 2002, half of it sagebrush communities (Wisdom et al., 2005). Besides direct fire suppression efforts, land managers have also sought to break this intensified fire cycle by rehabilitating native plant communities, primarily through direct seeding following wildfire (Monsen & Shaw, 2001). This reseeding strategy to restore native plant communities on western public rangelands, if successful, may be comparable to fire suppression in cost but will have more enduring habitat value.

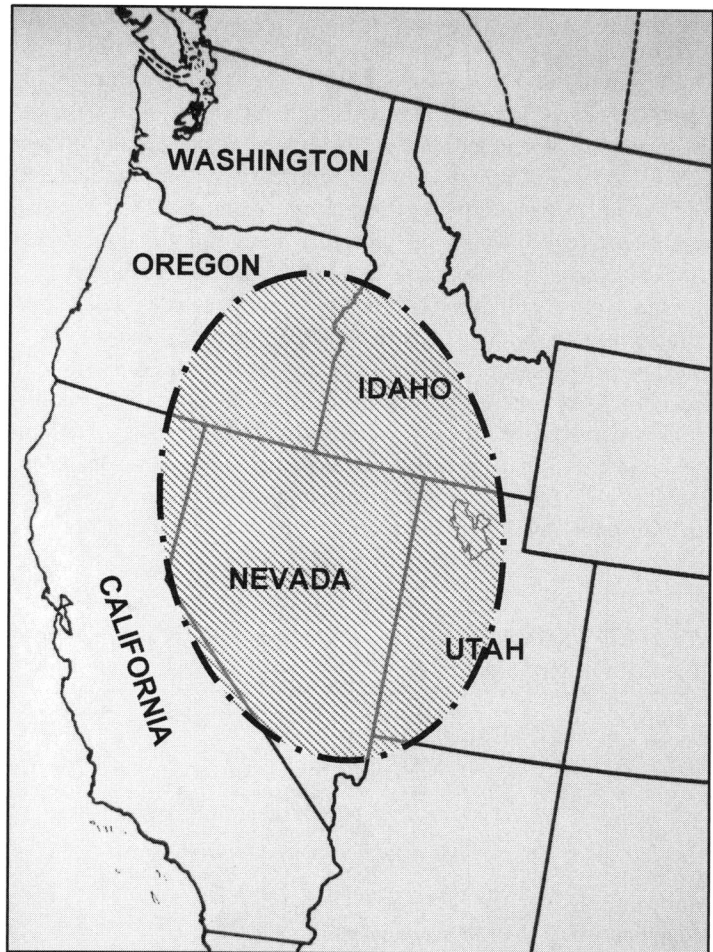

Figure 4.1 A map of the western U.S. Intermountain Region showing the hydrological Great Basin (shaded circle), which botanists often recognize as a single floristic region.

Recent Great Basin reseeding efforts to rehabilitate rangelands have been ambitious. However, native wildflowers are barely represented in these seed mixes, in terms of either volume or diversity, a circumstance that a new research program seeks to redress. The program scale has no precedent. From 1999–2004, massive wildfires burned across the western United States. In response, federal agencies distributed more than 6,500 metric tons of seed. Since the 1970s, the reseeding mixes have consisted of about one-third each of exotic grasses, native grasses, and native shrubs (Monsen & Shaw, 2001). Native flower seed constituted a mere 0.5% of the mix, most of that being a native yarrow, *Achillea millefolium* L. Dramatically more wildflower seed, about 150 tons each year, is sought by western land managers to rehabilitate the hundreds of thousands of hectares that annually burn in the Great Basin and adjacent biomes (Scott Lambert, National Seed Coordinator, Bureau of Land Management, 2007 personal communication). The sage grouse is hoped to be a major beneficiary; this endemic bird is otherwise fast becoming endangered. Native shrub seed for such restoration continues to be harvested directly from nature. Though cost effective for shrubs, wild harvest is impractical and unreliable for native wildflowers, as it yields paltry quantities of expensive seed. For example, wild-harvested seed of *Hedysarum boreale* was recently being marketed for US$110 per kilogram and that of *Sphaeralcea ambigua* for US$180 per kilogram. Budgetary constraints and harvest inefficiencies will forever limit the volumes of most wild-harvested wildflower seed that can be bought to satisfy these reseeding programs.

A cadre of innovative commercial growers are farming fields of the native grasses and, more recently, experimentally farming several of the native wildflower species desired for restoration seed. With the advent of the Great Basin Native Plant Selection and Increase Project in 2001, farming practices are being developed for 16 eudicot flowering species (table 4.1) native to the Intermountain Region, especially the Great Basin, but including the Columbia Plateau to the north and the Colorado Plateau to the

Table 4.1 Wildflower species desired for rangeland rehabilitation in the U.S. Intermountain West.

Family	Species
Apiaceae	*Lomatium dissectum* (Nutt.) Math. & Const.
	L. triternatum (Pursh) Coult. & Rose
Asteraceae	*Balsamorhiza sagittata* (Pursh) Nutt.
	Crepis acuminata Nutt.
Cleomaceae	*Cleome lutea* Hook
	C. serrulata Pursh
Fabaceae	*Astragalus filipes* Torr.
	Dalea ornatum (Dougl.) Barneby
	D. searlsiae (Gray) Barneby
	Hedysarum boreale Nutt.
	Lupinus argenteus Pursh
	L. sericeus Pursh
Malvaceae	*Sphaeralcea grossularifolia* (H. & A.) Rydb.
	S. munroana (Dougl.) Spach.
Plantaginaceae	*Penstemon speciosus* Dougl.
Polygonaceae	*Eriogonum umbellatum* Torr.

southeast. They were chosen in part for being regionally widespread, common within native plant communities, broadly adapted at mid- and lower elevations, tolerant of fire, and palatable to wildlife and for having practical promise for farming (e.g., mechanical harvest). By harnessing the efficiencies and reliability of modern agriculture, the hope is to dramatically boost seed production while slashing seed costs.

Applied Pollination Paradigms

For the successful farming of wildflower seed, each flowering crop's pollination needs must be understood and satisfied by practical means. Three of the four approaches to agricultural pollination are relevant and context-specific: (1) renting hives of managed honey bees, (2) managing nesting of nonsocial bees in provided substrates, and (3) practicing on-farm pollinator stewardship to favor unmanaged nesting populations of local wild bee communities. The fourth alternative, which is purchasing disposable hives of bumble bees, will likely prove impractical for reasons discussed herein. A novel mix of these pollination strategies is being developed for farmed wildflowers.

Applied pollination ecology is practiced in two traditional arenas, agriculture and wildflower conservation. Their unusual marriage in this program brings novel challenges and opportunities. In agriculture, extensive monocultures of food, forage, and fiber crops are farmed far beyond their geographic points of origin. Most pollinated crops are either herbaceous annuals or woody perennials, such as orchard fruits. Each crop's pollination needs and benefits are generally well understood (Free, 1993; Klein et al., 2007). Most crops are derived from decades, if not centuries, of artificial selection and sometimes controlled hybridization. Prolonged plant breeding often modified their reproductive biologies, inadvertently or intentionally eliminating physiological or mechanical barriers to selfing while favoring passive autopollination. The wild progenitors of a few crop species are extinct (e.g., onions, *Allium cepa*) and, with them, knowledge of their former natural pollinator faunas.

Farming Great Basin wildflowers for restoration seed contrasts sharply with farming our traditional crops, with profound consequences for pollination needs and pollinator management. Only two of the species are annuals; the rest are perennial, but herbaceous. They are being grown in or near their native geographic ranges. We are ignorant of their pollination needs and benefits, often for entire genera, sometimes even for tribes (Cane, 2005, 2006a) or families (e.g., Cleomaceae). The Association of Seed Certification Agencies (AOSCA) has promulgated new standards and protocols for native seed collection and certification called the "source identified class." It dictates ≤5 generations in cultivation before returning to geographically explicit wild seed sources. Artificial and inadvertent selection are being minimized and genetic diversity maintained in order to retain regional adaptations relevant for wildflower establishment once seeded back into wildlands. These objectives, strategies, and problems have no precedent in agriculture, although the tools and farming practices are mostly borrowed and adapted from traditional agriculture.

The second realm of applied pollination ecology addresses the breeding and pollination biology of wildflowers in a context of species conservation. Most of the funding focuses on specific threatened or endangered plant species growing in situ. Failed sexual reproduction has been a tenable explanatory hypothesis for rarity of endangered wildflowers, but more recent accumulating evidence implicates other factors, primarily habitat loss, for the endangerment of most continental flowering species that are pollinated by insects (Tepedino, 2000). Unlike crop plants, these rare wildflowers enjoy adequate natural pollination service, partly because their floral densities fall far short of those experienced in cropped monocultures, and so the densities of unmanaged wild pollinators typically suffice. Many ecologically important plant genera lack such endangered species (or they have not been studied). Consequently, studies of endangered wildflowers leave us surprisingly ignorant of the breeding biologies and pollinators of the numerous native wildflowers that dominate plant communities targeted for ecological restoration (Cane, 2006a), including those of the Great Basin. Thus neither agricultural nor conservation contexts for applied pollination research provide a satisfactory blueprint to address pollination needs of farmed native wildflowers whose harvested seed will later be used to rehabilitate and restore native plant communities.

Pollination Needs

None of these Great Basin wildflowers that are under study in this project is wind pollinated, and none proliferates vegetatively. Only one species, *Crepis acuminata*, can set considerable seed without pollinators; most of its populations include numerous polyploid individuals that can reproduce asexually through apomixis (Babcock & Stebbins, 1938). Even for the more self-fertile species, comparatively little seed results from autopollination (unaided pollen transfer; figure 4.2). The two *Cleome* species constitute a partial exception, as mechanical jostling of densely grown flowering stands nearly doubled fruit set in cages (37% vs. 19% (Cane, 2008)). Nevertheless, all nine species studied at my laboratory have produced dramatically more seed with pollinator visitation (see figure 4.2). Northern sweetvetch (*Hedysarum boreale*) sets no fruit or seed at all if pollinators are excluded from its flowers (Swoboda, 2007). The species also vary in their degrees of self-fertility, a trait useful for colonization. Outcrossing has yielded more modest gains in seed or fruit production (figure 4.3). These gains are nonetheless substantial. For example, outcrossing by *Astragalus filipes* doubles seed yields and could thereby halve seed prices from the same field. For a few cases, such as *H. boreale*, flowers pollinated freely by bees set more seed than even our manual outcrossing treatments, suggesting further nuances for maximal seed set (Swoboda, 2007). Overall, if literally tons of wildflower seed will be produced cheaply and profitably, then fruit and seed set must be maximized through judicious use of reliably abundant and effective pollinators.

Pollinator guilds of each of these wildflower species consist exclusively of insects. For 12 species, I have systematically sampled 1–6 floral visitor guilds in 1–3 states, along with any bees in our experimental common garden in Logan, Utah. Native bees, sometimes joined by honey bees, dominate most of the guilds, with up to 29 bee species represented in pooled guild samples of no more than 200 individuals. The small floral visitor

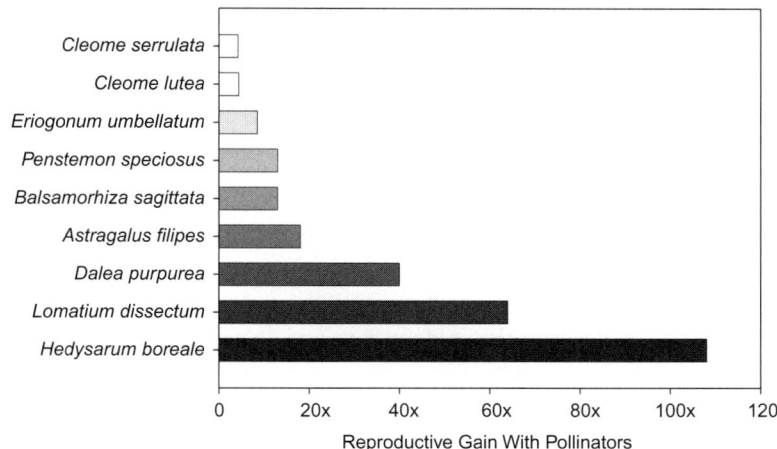

Figure 4.2 Western wildflowers grown for rangeland restoration and their requirements for pollinators. Using seed yield per inflorescence, reproductive gain was calculated as the average yield of seeds from flowers open to pollinator visits, divided by the yield from flowers bagged in mesh netting that excluded pollinators but allowed wind pollination and autopollination. Treatments were either paired within plants or randomly assigned among plants. Pollinator numbers were not supplemented. Note that the abscissa is a multiplier, not a percentage.

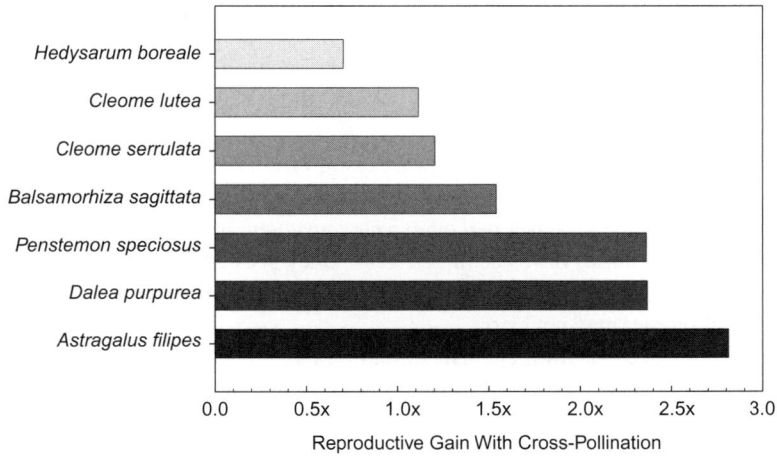

Figure 4.3 Self-fertility of western wildflowers grown for rangeland restoration in the Western United States. Flowers netted or caged to exclude insect visits were manually pollinated with pollen from either the same or a different plant. Plants (8–25 per species) were grouped for each replicate treatment.

guild at *Penstemon speciosus* constitutes one exception. Its two specialist *Osmia* bees are joined by another prominent visitor and likely pollinator, the pollen wasp, *Pseudomasaris vespoides* (Masaridae), which frequents flowers of this genus (Tepedino, 1979). The two *Cleome* species are again exceptional, too, their prodigious and accessible nectar attracting diverse nectarivorous insects, primarily butterflies, bees, and wasps. Their wild populations have yet to be systematically sampled. Curiously, although these *Cleome* flowers are visited and readily pollinated diurnally, I have found that their anthers dehisce and stigmas become receptive nocturnally (Cane, 2008). In desert basins (but not at the common garden) their flowers might also be visited by moths. The other wildflowers are all diurnally pollinated. Overall, it has become clear that each of these wildflower species needs pollinators. Bees are the predominant and diverse group of visitors populating nearly all of their floral guilds. Bees seem likely, therefore, to be generally essential for these wildflowers' seed production.

Pollinator Options

Honey Bees

The European honey bee (*Apis mellifera*) continues to be the versatile workhorse of agricultural pollinators. In many developed countries, hives of honey bees are typically rented from professional migratory beekeepers who transport palletized hives to pollinate sequentially flowering crops. Honey bees are, therefore, the obvious default choice to pollinate farm fields of wildflowers, too. Even at a rental cost of US$100 per hive, a strong colony's 10,000 foragers are cheap, each rented forager costing the grower 1 cent. As an added advantage, once a beekeeper is contracted and arrives with rental hives, a grower has no further pollination inputs or concerns.

But honey bees do have drawbacks, some of which are specific to the practicalities of wildflower seed production:

1. Some of today's native seed growers are distant from regions in which migratory beekeepers pollinate crops. Delivering small numbers of rental hives to a small, out-of-the-way wildflower grower will often be inconvenient at best.
2. Mite-borne diseases have decimated feral honey bees and thinned the ranks of hobbyist beekeepers, while doubling costs for commercial hives. Hence, feral or locally owned colonies may not exist.
3. For certain seasons and regions, pollination demand exceeds hive availability, most notably in the early spring when tree fruits and almonds bloom in the western United States.
4. Field size for a given wildflower seed crop is likely to remain small by some agricultural standards. If a given cultivated wildflower species has but mediocre attraction for foraging honey bees, their broad floral preferences and flight vagility will allow them easy access to alternative floral hosts in neighboring fields, fallow areas, or surrounding wild lands.

5. Lastly, honey bees may be ineffective pollinators for a given wildflower species. For these reasons, honey bees will be impractical pollinators for some native wildflower seed crops.

For a few traditional crops, nontraditional bees are superior pollinators and can be effectively managed in the large numbers that are needed. Sometimes these alternative pollinators are useful only in particular cultural or economic contexts for a given crop, such as the use of *Osmia cornifrons* (Megachilidae) to pollinate apples only in Japan, or the use of several *Bombus* species whose commercial colonies are shipped great distances to pollinate tomatoes in North America, Europe, and the Mediterranean, but only in greenhouses (see chapters 3 and 9, this volume). Whatever the pollinator, economics and practical considerations for managing the bees and the crop constrain our options for reliably supplying large numbers of dedicated pollinators to a given crop monoculture.

Bumble Bees

My bee surveys revealed bumble bees to be often ubiquitous and sometimes common members of these wildflowers' pollinator guilds. They avidly visit and effectively pollinate two of the four species of legumes, *Astragalus filipes* and *Hedysarum boreale*. In our common garden and in trial cultivation plots, unmanaged bumble bees also have proven to be excellent pollinators of *Penstemon speciosus*, although they were not seen at this *Penstemon* in the wild. Stewardship of unmanaged bumble bee populations in and around native seed farms (see chapter 2, this volume) might therefore pay dividends in seed yields to growers. However, commercial bumble bee colonies are costly for open-field pollination, with individual foragers costing about one hundredfold more than rented honey bees. Purchased bumble bee colonies are short-lived, shifting production from workers to drones and queens after a few months. Hence, they are disposable. For now, they are economically practical only when foragers pollinate highly valuable crops in confinement, such as greenhouses (see chapter 3, this volume). Off-season or early season fruit production in greenhouses gives growers profitable marketing advantages that justify the costs of disposable bumble bee colonies. However, there is no incentive in being the first grower of the year with fresh wildflower seed, and so no justification exists for greenhouse cultivation and its attendant bumble bee colony purchases. For native seed crops flowering in open fields during the growing season, pollinators less expensive than commercial bumble bees should be able to satisfy pollination needs.

Currently Managed Nonsocial Bees

Some managed but nonsocial bees are likely to find use in pollinating wildflower seed crops. In North America, farm management protocols exist for three species. The alfalfa leafcutting bee (*Megachile rotundata*) and the blue orchard bee (*Osmia lignaria*) nest above ground in cavities, whereas the alkali bee (*Nomia melanderi*) nests underground. As with any bee, nesting needs, flight seasonality, floral host preferences, and pollination efficacy together will define their promise as pollinators of a native wildflower species for seed production. The question then remains: Can we sustainably manage these bees in large enough numbers to adequately stock wildflower production fields?

Management methods and markets are mature for the alfalfa leafcutting bee (see chapter 7, this volume); this bee has revolutionized alfalfa seed production over the past 40 years (Stephen, 2003). It does require summer's warmth for development and flight, precluding its use for pollinating springtime wildflowers. It is particularly adept at pollinating small-flowered legumes such as alfalfa. The prairie clovers (*Dalea*) are also small-flowered, summer-blooming, herbaceous perennial legumes, two of which are desired for Great Basin reseeding (see table 4.1). A third (prairie) species, *Dalea purpurea*, was commercially available for immediate transplant into our outdoor trial garden, and so I used it as a surrogate for the Great Basin species that were being grown from seed. As expected, the alfalfa leafcutting bee was found to freely forage at its uncaged flowers, choosing to provision its nest cells solely with *D. purpurea* pollen. Stigmatic contact appeared inevitable because the bees scrabbled for pollen on the dense inflorescences. I obtained substantial seed sets per plant (>20,000 seeds, and up to 80% seed set; Cane, 2006a).

Honey bees, bumble bees, and several wild solitary bee species were also seen visiting these *Dalea* flowers concurrently with the alfalfa leafcutting bee, so the leafcutting bee's specific contribution to pollination awaits assessment. In cages, at least, alfalfa leafcutting bees are useful in cross-pollinating small-flowered specialty seed crops (e.g., carrots; Tepedino, 1997) or hybrid foundation seed or in regenerating small lots of seed for seed repositories, showing more versatility than might be expected from their floral foraging preferences. This leafcutting bee could likewise be flown in field cages to bulk up small lots of seeds harvested from wild stands of forbs. This could yield the several kilos of clean foundation seed to be later sown by growers (*Lomatium* is an apt candidate, also being of the carrot family). The field pollination needs of several of the summer-blooming, small-flowered wildflowers such as *Dalea* spp. could be satisfied by alfalfa leafcutting bees as well, managed in the same way as for fields of seed alfalfa (see chapter 7).

Ground-Nesting Bees of Agriculture

The pollinator guilds of all of the target wildflower species I have surveyed thus far include multiple species of ground-nesting bees, sometimes to the exclusion of any other kind (e.g., for *Lomatium dissectum*). Numerous other ground-nesting bees effectively pollinate various crop species as well (Cane, 1997), but only the alkali bee, *Nomia melanderi*, is intensively managed (Johansen et al., 1978). Like the alfalfa leafcutting bee, it flies soon after midsummer and is used to pollinate seed alfalfa (Cane, 2002; see also chapter 7, this volume). It may prove useful to pollinate species of *Dalea* and *Cleome*; *C. lutea* is suspected to be a key natural floral host of the alkali bee (Richard Rust, personal communication, 2006).

Unlike the alfalfa leafcutting bee, the alkali bee presents daunting nest establishment challenges typical of other ground-nesting bees. It requires silty subirrigated soils with salty surfaces. Managed aggregations can be densely populous (more than 5 million females) and can endure for 50 years or more (Cane, 2008). However, where naturally silty, subirrigated soils are absent, these earthen nesting sites can be difficult and expensive to construct. Nest sites are then populated by coring, transporting, and inserting hundreds or even thousands of large 30 kg soil blocks bearing overwintering prepupae. The blocks are punched from the parent aggregation using custom hydraulic tools. Native seed crops might be profitably produced adjacent to managed alkali bee aggregations, a novel strategy of bringing the crop to the bee rather than vice versa. However, unless wildflower seed growers have the long-term commitment of many alfalfa seed growers to maintain nesting populations of the alkali bee, active management of this and other ground-nesting bee species is impractical or impossible for agricultural pollination, including wildflower seed production.

In only a few cases have wild but unmanaged ground-nesting bees proven to be both effective and adequately abundant across the entire range of cultivation for any one crop. In the southeastern United States, the southeastern blueberry bee (*Habropoda laboriosa*; Apidae) and bumble bees are together responsible for pollinating farmed rabbiteye blueberries (*Vaccinium ashei*) on all but the largest farms (Cane & Payne, 1993). At some North American gardens and commercial fields of squashes and pumpkins (*Cucurbita* spp.), the squash bees (esp. *Peponapis pruinosa*; Apidae) satisfy pollination needs (Hurd et al., 1971; Tepedino, 1981; Roulston et al., 1996; Shuler et al. 2005). A collaborative survey of pollinators amid cultivated squashes and pumpkins is demonstrating the predominant pollination service of *Peponapis* bees throughout most of the Western Hemisphere (Cane et al., unpublished data). Nesting populations of *P. pruinosa* can apparently increase enough to pollinate cultivated squashes and pumpkins at field sizes up to at least 75 ha. (R. Hammon, personal communication, 2006). Both *H. laboriosa* and *P. pruinosa* are ground-nesting floral specialists that effectively pollinate their respective crops (Tepedino, 1981; Sampson & Cane, 2000). Importantly, both crops are typically grown at a location for years at a time, accommodating the slow growth of pollinator populations detailed subsequently.

Untried Bees to Manage

Potentially manageable, effective pollinators exist among North America's 4,000 nonsocial bee species. Most promising are the 139 species of North American *Osmia*, for half of which species we already know basic nesting habits and many of which are cavity nesters (Cane et al., 2007). Some have proven themselves amenable to artificial nesting substrates to pollinate various perennial crops (e.g., apples, raspberries, almonds; see, e.g., Bosch, 1995; Torchio, 2003; Cane, 2005, 2006b). Unlike species of *Megachile* or *Hoplitis*, which overwinter as prepupae, all *Osmia* overwinter as adults. As a group, *Osmia* are therefore more phenologically versatile, some species emerging with the first warm days of spring (e.g., *O. lignaria*; see chapter 6, this volume). Others, such as *O. bruneri* (Frohlich, 1983) or *O. sanrafaelae* (Parker, 1985), require days of warm incubation that delay their emergence until midsummer. To match on-farm bloom phenology, emergence date for a given *Osmia* species can be manipulated somewhat to shift it earlier or later by several weeks. Alternatively, populations can be sourced from places with cooler or warmer climates to better match a seed crop's local blooming schedule. If a wildflower is farmed for seed where its natural bloom time is advanced or retarded, novel matches of bee and wildflower are possible. One or more species of *Osmia* are usually present and sometimes prevalent in the guilds of floral visitors that I have thus far sampled at the wildflowers targeted by this project (e.g., dominating every one of 21 *Astragalus filipes* populations sampled in four states). Some of these cavity-nesting *Osmia* occur in more than one floral guild. For instance, I have sampled *O. bruneri* at three of the four legumes (*Hedysarum boreale*, *Astragalus filipes* and *Lupinus sericeus*). Such floral versatility foreshadows managing one *Osmia* bee species for several different flowering crops.

Unmanageable Bees to Try

Unmanaged wild bees pollinate these wildflowers in nature, so why not rely on this strategy for farmed wildflower seed crops as well? In wildflower fields of several native seed growers, I have found ground-nesting native bees foraging and nesting, albeit in sparse numbers. In home and market gardens, ratios of bloom to nearby nesting opportunities can yield satisfactory densities of such wild pollinators. However, at larger commercial field sizes, wild pollinator numbers typically become greatly diluted (e.g., Scott-Dupree & Winston, 1987), hence the need to supplement with managed pollinators. To be profitable, most family and corporate farms of the developed world must depend on either mechanization and economies of scale or niche marketing. Farmed flowering monocultures present a "sea of bloom" to bees, with plentiful pollen and nectar rewards. Insufficient undisrupted nesting opportunities may limit growth and ultimate size of nonsocial bee populations, judging from the inadequate pollination service provided by unmanaged bees in larger fields and orchards.

Nesting opportunities for wild bees are often restricted to the field perimeter, which likely explains their sparsity in the larger fields and orchards of commercial agriculture. When nesting is restricted to field margins, simple edge:area relationships dictate that nesting densities must double for every fourfold increase in field size to maintain constant floral visitation intensity in the field. The predicament is obvious for those nonsocial cavity-nesting bee species that nest in abandoned beetle burrows in dead wood or in pithy dead twigs or stems. These nesting opportunities are scarce or nonexistent within larger fields and orchards in many farming regions. Bumble bees may fare no better, because most species nest either above ground in tree hollows or underground in abandoned nests of rodents or other small mammals.

Even for ground-nesting nonsocial bee species, fields and orchards rarely offer sufficient nesting habitats to match the prodigious pollination needs of traditional crops. These constraints likely foreshadow similar problems for wildflower seed growers who expect to depend on unmanaged pollinators. Most ground-nesting species seek patches of exposed, sunlit bare soil, sometimes as vertical embankments. Farming practices and crops themselves often limit these opportunities, owing to impenetrable shady canopies (e.g., seed alfalfa, cranberries), impermeable plastic mulches (e.g., commercial strawberries), or dense interalley turf (e.g., tree fruit orchards, blueberries). Even for crops such as sunflower with suitably exposed soil (Wuellner, 1999), controlling weeds during bloom by interrow cultivation buries these bees' nest entrances. Deeper nesting species escape tillage damage once their nesting season is done. Not so for bees nesting shallowly in the soil, whose nests are destroyed by cultivation. Their stewardship may require year-round no-till practices. These shallow nesters include all of the species of *Megachile* (Eickwort et al., 1981) and *Osmia* (Cane et al., 2007) that nest shallowly underground, which are the very species I have sampled from pollinator guilds of *Astragalus, Crepis, Hedysarum, Lupinus,* and *Penstemon* (see table 4.1, figure 4.2). This group of ground-nesting bees also includes effective pollinators of traditional crops (Hobbs, 1956; Cane et al., 1996). Whether or not they can ultimately satisfy the pollination needs of native seed growers remains to be seen.

Growers will need patience with the realized reproductive outputs of unmanaged solitary bees; they multiply slowly for an insect. A female of a nonsocial species starts life with 30 or so eggs. Half or more of these will yield males. Females of most species will typically provision only one or two nest cells daily during good weather. Larval diseases and nest parasites and predators can subtract substantially from a mother bee's reproductive output. Hence, in the first years that a crop blooms, starting numbers of unmanaged bees will fall far short of densities needed to satisfy the sudden pulse in pollination need. Numbers of unmanaged pollinators will likely remain inadequate for some years while their populations gradually grow. Overstocking of managed species, such as honey bees, may further retard this growth through floral resource competition. Wild bees may initially be more populous if availed another adjacent, usable flowering crop in the preceding years. The circumstances on most wildflower seed farms tend to match those found to favor pollination of traditional crops by unmanaged wild bees: fields adjacent to wildlands, small field size, little insecticide use, and either alternate floral resources nearby (for floral generalists) or a perennially grown crop that blooms year after year (for its floral specialists). Even under these conditions, adequately abundant managed pollinators will be needed initially to pollinate the crop, serving as a bridge between the first years of a new wildflower crop and the year when the numbers of wild bees finally grow to satisfy pollination needs. If and when patient stewardship results in sufficiently populous ground-nesting pollinators to service bloom, they might be the least expensive and simplest means of pollinating a given wildflower crop.

Flow of Research

Research for this program must progress simultaneously on multiple fronts to keep apace of grower needs (Bosch & Kemp, 2002). I am following a logical sequence of steps:

1. Plots of source-identified wildflowers are needed in our common garden for later caged pollination trials. Using wild populations in situ is logistically impractical if they are separated by hundreds of kilometers. Manual pollination treatments and later seed harvest can

require daily attention, with intervening protection from herbivores, seed predators, and a curious public (Cane, 2005). Because most of these wildflowers are unavailable from nurseries, my collaborators collect seed from the wild, which are then germinated and grown in the greenhouse. Once transplanted outside, I await first flowering, typically 1–3 years later for these perennials.

2. To characterize the small subset of potential pollinators from each bee community, I systematically net sample guilds of each wildflower's floral visitors throughout their geographic ranges. The locations of floral host populations and estimated bloom dates are gleaned from specimens housed in the region's herbariums.

3. To obtain starting populations of the subset of potentially manageable cavity-nesting species, both drilled nesting blocks with multiple-sized holes and drilled stick nests are manufactured and deployed. Most are placed in the wild, preferably in the presence of target wildflower hosts, but we have enjoyed some success placing them around grower's wildflower fields, too. With luck, some of the cavity-nesting species thus obtained match those surveyed at a wildflower's visitor guild. Each nest's progeny are identified by sacrificing only one individual, preferably a male. This leaves the rest to fly, forage, and nest the following year in confinement with one of the common garden stands of wildflowers.

4. Their prowess as pollinators is assessed against our own manual pollination treatments from the preceding year.

5. Stocking densities of the bees are estimated that give both abundant seed yields and sustainable population growth of the bees when limited to a particular floral species for forage.

Two case studies show that success proceeds apace of progress with each of these components while overcoming idiosyncrasies of bees, bloom, and weather. For *Balsamorhiza sagittata*, which is estimated to take more than 5 years from seed to bloom, I experimented with the species' breeding biology using nearby wild populations (Cane, 2005). It was found to require insect pollinators (see figure 4.2) and to benefit from outcrossing (see figure 4.3). Two widespread cavity-nesting bees, *O. californica* and *O. montana*, were commonly sampled visiting its flowers. Both were earlier reported to forage at *Balsamorhiza* and *Wyethia* (Torchio, 1989). When collecting pollen, they were seen to vigorously pat pollen-bearing surfaces of the stigmatic lobes using their ventral abdominal scopae, thereby inevitably pollinating as well. Populations readily increased using my practical nesting system (Cane, 2006b). An estimated stocking density of several hundred females per hectare satisfies pollination needs (Cane, 2005). These two *Osmia* species answer for all of the objectives of the program and are ready to manage as pollinators for seed production of any *Balsamorhiza* species.

I have pursued a different approach and a different path to pollinator management with my studies of *Dalea* (Cane, 2006a). To accelerate my research schedule, I obtained transplants of a Great Plains species, *D. purpurea*, for our common garden in Logan. It served as a surrogate species while I awaited seed and maturation of the two Great Basin species, neither of which has been previously cultivated nor is found within 250 km of Logan. Pollination trials in our common garden revealed a necessity for pollinators (see figure 4.2) and great benefits from outcrossing (figure 4.3). Its rich Midwestern pollinator guild suggests many choices for pollinators, a guess borne out in our common garden, where honey bees, bumble bees, alfalfa leafcutting bees, and various wild species eagerly sought its flowers. Fertile seeds formed at 60–80% of available flowers. A small, managed nesting population of alfalfa leafcutting bees preferentially provisioned numerous nest cells with the bright orange pollen of *D. purpurea* flowers. Because inflorescences of the two Great Basin *Dalea* spp. closely resemble those of *D. purpurea*, I am optimistic that either honey bees or managed alfalfa leafcutting bees will handily pollinate farmed Great Basin *Dalea*. Initially small populations of wild bees seen visiting newly planted *Dalea* could multiply in subsequent years in response to simple stewardship practices, ultimately satisfying the pollination needs of farmed *Dalea*.

Summary

Federal, state, and private land managers in the western United States annually seek tons of affordable seed from dominant native wildflowers for use in their wildland rehabilitation and restoration programs. The prohibitive price of wild-collected wildflower seed stymies that effort. Seed costs and practicality necessitate the farming of wildflowers for their seed. Heterogeneous cultivation needs, seed harvest methods and site-specific restoration seed mixes all dictate that farmers grow these species in monocultures rather than meadow mixes. However, individual native seed growers can be expected to be farming more wildflower species as their skill and success meet demands.

I am finding that bees are needed to pollinate nearly all of the target wildflower species desired for restoration and rehabilitation in Great Basin plant communities. Actively managed bees will initially be necessary, at least as a "bridge" for growers in early production years. If and when wild bees proliferate to become abundant on farms, then some farms and seed crops will eventually enjoy substantial pollination services from passively managed local native bee communities. Context, practicality, and cost will determine the mix of bees used. Options under study for actively managed pollinators include honey bees, currently managed nonsocial species (*M. rotundata*, *O. lignaria*), and additional cavity-nesting species of *Osmia*, *Megachile* and *Hoplitis* that show pollination prowess and nest management potential. Native bees will need to be sustainably managed on seed farms, as it is impractical and costly to replenish managed populations by annually trap-nesting wild bee populations. Unmanaged farm populations of bumble bees and ground-nesting bees should benefit from some simple pollinator stewardship practices, including a hiatus in cultivation during bee nesting, wise insecticide use, and planting supplemental forage crops nearby.

Pollinator lessons learned with the Great Basin Plant Selection and Increase Project should translate to other regions and biomes in which native wildflower seed is increasingly sought for restoration around the United States, including the Upper Colorado Plateau, the Mojave Desert, and the Great Plains. Successful seeding and restoration of native wildflowers will have far-reaching positive ecological impacts on damaged Great

Basin plant communities and their dependent native bee communities. These gains will dwarf the comparatively small economic impact of a peculiar native seed industry and the novel mix of bees that it will use to produce its wildflower seed.

References

Babcock, E. B., & Stebbins, G. L. (1938). *The American species of* Crepis: *Their interrelationships and distribution as affected by polyploidy* (Carnegie Institution Publication no. 504). Washington, DC: Carnegie Institution.

Bosch, J. (1995). Comparison of nesting materials for the orchard pollinator *Osmia cornuta* (Hymenoptera: Megachilidae). *Entomologia Generalis, 19,* 285–289.

Bosch, J., & Kemp, W. P. (2002). Developing and establishing bee species as crop pollinators: The example of *Osmia* spp. (Hymenoptera: Megachilidae) and fruit trees. *Bulletin of Entomological Research, 92,* 3–16.

Cane, J. H. (1997). Ground-nesting bees: The neglected pollinator resource for agriculture. *Acta Horticulturae, 437,* 309–324.

———. (2002). Pollinating bees (Hymenoptera: Apiformes) of U.S. alfalfa compared for rates of pod and seed set. *Journal of Economic Entomology, 95,* 22–27.

———. (2005). Pollination needs of arrowleaf balsamroot, *Balsamorhiza sagittata* (Heliantheae: Asteraceae). *Western North American Naturalist, 65,* 359–364.

———. (2006a). An evaluation of pollination mechanisms for purple prairie-clover, *Dalea purpurea* (Fabaceae: Amorpheae). *American Midland Naturalist, 156,* 193–197.

———. (2006b). The Logan BeeMail shelter: A practical, portable unit for managing cavity-nesting agricultural pollinators. *American Bee Journal, 146,* 611–613.

———. (2008). Breeding biologies, pollinating bees and seed production of *Cleome lutea* and *C. serrulata* (Cleomaceae). *Plant Species Biology,* in press.

———. (2008) A native ground-nesting bee (*Nomia melanderi*) sustainably managed to pollinate alfalfa across an intensively agricultural landscape. *Apidologie,* in press.

Cane, J. H., Griswold, T., & Parker, F. D. (2007). Substrates and materials used for nesting by North American *Osmia* bees (Hymenoptera: Apiformes: Megachilidae). *Annals of the Entomological Society of America, 100,* 350–358.

Cane, J. H., & Payne, J. A. (1993). Regional, annual and seasonal variation in pollinator guilds: Intrinsic traits of bees (Hymenoptera: Apoidea) underlie their patterns of abundance at *Vaccinium ashei* (Ericaceae). *Annals of the Entomological Society of America, 86,* 577–588.

Cane, J. H., Schiffhauer, D., & Kervin, L. J. (1996). Pollination, foraging, and nesting ecology of the leaf-cutting bee *Megachile* (*Delomegachile*) *addenda* (Hymenoptera: Megachilidae) on cranberry beds. *Annals of the Entomological Society of America, 89,* 361–367.

Eickwort, G. C., Matthews, R. W., & Carpenter, J. (1981). Observations on the nesting behavior of *Megachile rubi* and *M. texana* with a discussion of the significance of soil nesting in the evolution of megachilid bees (Hymenoptera: Megachilidae). *Journal of the Kansas Entomological Society, 54,* 557–570.

Free, J. B. (1993). *Insect pollination of crops.* New York: Academic Press.

Frohlich, D. R. (1983). On the nesting biology of *Osmia* (*Chenosmia*) *bruneri* (Hymenoptera: Megachilidae). *Journal of the Kansas Entomological Society, 56,* 123–130.

Hobbs, G. A. (1956). Ecology of the leaf-cutter bee *Megachile perihirta* Ckll. (Hymenoptera: Megachilidae) in relation to production of alfalfa seed. *Canadian Entomologist, 87,* 625–631.

Hurd, P. D., Jr., Linsley, E. G., & Whitaker, T. W. (1971). Squash and gourd bees (*Peponapis, Xenoglossa*) and the origin of the cultivated *Cucurbita*. *Evolution, 25,* 218–234.

Johansen, C. A., Mayer, D. F., & Eves, J. D. (1978). Biology and management of the alkali bee, *Nomia melanderi* Cockerell (Hymenoptera: Halictidae). *Melanderia, 28,* 25–46.

Klein, A. M., Vassiere, B. E., Cane, J., Steffan-Dewenter, I., Cunningham, S., Kremen, C., et al. (2007). Importance of pollinators in changing landscapes for world crops. *Proceedings of the Royal Society of London: Series B, 274:* 303–313.

Monsen, S. B., & Shaw, N. L. (2001). Development and use of plant resources for western wildlands. *Proceedings of the Rocky Mountain Research Station P-21.* Ogden, UT: USDA Forest Service, 47–61.

Parker, F. D. (1985). A candidate legume pollinator, *Osmia sanrafaelae* Parker (Hymenoptera: Megachilidae). *Journal of Apicultural Research, 24,* 132–136.

Roulston, T., Sampson, B., & Cane, J. H. (1996). Squash and pumpkin pollinators plentiful in Alabama. *Alabama Agricultural Experiment Station, 43,*19–20.

Sampson, B. J., & Cane, J. H. (2000). Pollination efficiencies of three bee (Hymenoptera: Apoidea) species visiting rabbiteye blueberry. *Journal of Economic Entomology, 93,* 1726–1731.

Scott-Dupree, C. D., & Winston, M. L. (1987). Wild bee pollinator diversity and abundance in orchard and uncultivated habitats in the Okanagan Valley, British Columbia. *Canadian Entomologist, 119,* 735–745.

Shuler, R. E., Roulston, T. H., & Farris, G. E. (2005). Farming practices influence wild pollinator populations on squash and pumpkin. *Journal of Economic Entomology, 98,* 790–795.

Stephen, W. P. (2003). Solitary bees in North American agriculture: A perspective. In K. Strickler & J. H. Cane (Eds.), *For nonnative crops, whence pollinators of the future?* (41–66). Lanham, MD: Entomological Society of America.

Swoboda, K. A. (2007). The pollination ecology of *Hedysarum boreale* Nutt. and evaluation of its pollinating bees for restoration seed production. Unpublished master's thesis, Utah State University.

Tepedino, V. J. (1979). Notes on the flower-visiting habits of *Pseudomasaris vespoides* (Hymenoptera: Masaridae). *Southwestern Naturalist, 24,* 380–381.

———. (1981). The pollination efficiency of the squash bee (*Peponapis pruinosa*) and the honey bee (*Apis mellifera*) on summer squash (*Cucurbita pepo*). *Journal of the Kansas Entomological Society, 54,* 359–377.

———. (1997). A comparison of the alfalfa leafcutting bee (*Megachile rotundata*) and the honey bee (*Apis mellifera*) as pollinators for hybrid carrot seed in field cages. In K. W. Richards (ed.) *Acta Horticulturae 437*: 457–461.

———. (2000). The reproductive biology of rare rangeland plants and their vulnerability to insecticides (USDA APHIS Technical Bulletin No. 1809: Grasshopper Integrated Pest Management Users Handbook). Riverdale, MD: USDA Animal and Plant Health Inspection Service.

Torchio, P. F. (1989). In-nest biologies and development of immature stages of three *Osmia* species (Hymenoptera: Megachilidae). *Annals of the Entomological Society of America, 82,* 599–615.

Torchio, P. F. (2003). The development of *Osmia lignaria* Say (Hymenoptera: Megachilidae) as a managed pollinator of apple and almond crops: A case study. In K. Strickler & J. H. Cane (Eds.), *For nonnative crops, whence pollinators of the future?* (67–84). Lanham, MD: Entomological Society of America.

Wisdom, M. J., Rowland, M. M., & Surber, J. L. (2005). *Habitat threats in the sagebrush ecosystem: Methods of regional assessment and applications in the Great Basin.* Lawrence, KS: Alliance Communications Group.

Wuellner, C. T. (1999). Nest site preference and success in a gregarious, ground-nesting bee *Dieunomia triangulifera. Ecological Entomology, 24,* 471–479.

5 Honey Bees, Bumble Bees, and Biocontrol

New Alliances Between Old Friends

Peter G. Kevan, Jean-Pierre Kapongo, Mohammad Al-mazra'awi, and Les Shipp

Introduction

Bees are well known for their ability to carry microscopic particles. After all, they are important pollinators, and most pollen is microscopic (Wodehouse, 1959). Bees also are known to carry fungal spores and bacterial cells, some of which are pathogens of the bees themselves, or of plants (Morse & Nowogrodzki, 1990; Shaw, 1999). The capacity of bees to vector spores, bacteria, and viruses can be turned to our advantage by using them to transport biological control agents, a technique known as pollinator biocontrol vector technology (Kevan et al., 2001, 2003, 2004, 2005). This chapter describes the range of biocontrol agents useful for this technique and the targets against which they can be used. These biocontrol agents all have good potential for control of weeds, plant pathogens, or insect pests. The safety of the vector is considered next, along with the fact that it must be harmonized with efficacy against the target problem. Following from and closely linked to those two parts of the pollinator biocontrol vector technology is the matter of concentration of the biocontrol agent in an effective formulation and the efficiency of its acquisition and delivery by the vector. Dispenser designs are discussed as they influence the efficiency of pickup of the formulated biocontrol agent by the pollinators and so constitute another component of the technology. Finally, we briefly consider environmental safety and issues of nontarget effects.

The elements involved in this new technology are:

1. A crop in need of protection
2. A pest (weed, disease, or herbivore) that adversely affects crop production
3. Pollinating insects that visit the flowers of that crop

4. Biocontrol agents that can be vectored by the pollinators
5. A formulation of that biocontrol agent that allows effective vectoring without unduly risking the health of the vectors but that also has demonstratable efficacy against the pest
6. Some means of dosing the vector (i.e., a dispenser)
7. Consumer and environmental safety

Even though this technology has been developed to manage plant diseases and pests, successful application requires interdisciplinary research on several fronts. The biocontrol agent should be effective against the target, yet relatively safe for the bee vectors. The agent must be dispersed by the bees in sufficient quantity to be effective against the target, yet the crop in need of protection must not be adversely affected. Plant and insect pathology must be part of research and development because microbial control agents are being used. Design of the inoculum dispensers requires insights into vector behavior, as does the technology of formulation to maximize inoculum dispersibility and dosage at the target site without negatively endangering the health of the vectors. Moreover, safety of nontarget organisms in the environment in which the technology is applied must be considered. Finally, any produce grown and protected through use of the pollinator biocontrol vector technology must be safe for consumption by people or livestock. Figure 5.1 illustrates the integrated and connected array of facets of the research and development process.

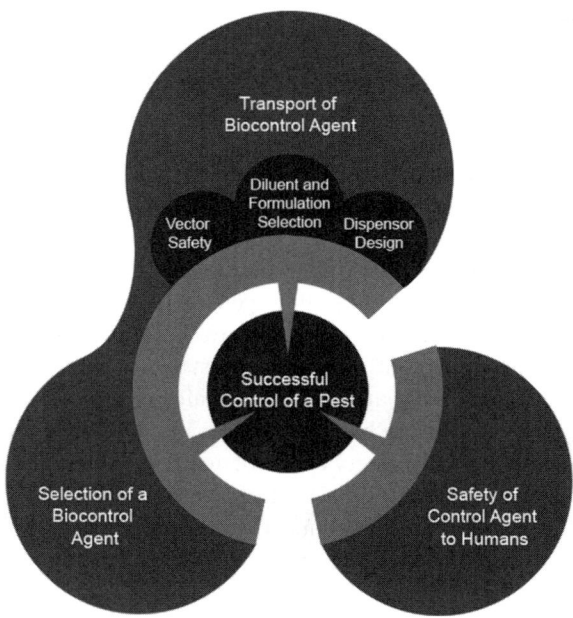

Figure 5.1 The integrated and connected array of facets of the research and development process in pollinator biocontrol vector technology.

Pollinator biocontrol vector technology has been explored by different investigators in a wide variety of ways with different biocontrol agents in various formulations, using different insect pollinators and different dispensing devices. Because few investigations report on comparisons and those that do report on only on a few in any one study, this chapter cannot provide an objective and comprehensive comparison of details. Readers should be aware that pollinator biocontrol vector technology has great potential in crop protection, and those interested in using it in their own research should refer to the cited papers in deciding how they should proceed. A great deal remains to be done to discover how pollinator biocontrol vector technology should be customized to the problems at hand.

Control Agents

Control of Seed Set in Weeds and Invasive Plants

One of the first considerations for using this technology was to control the fruiting and seed production of field milkweed, *Asclepias syriaca* (Asclepiadaceae; Eisikowitch et al., 1990; Kevan et al., 1989a, 1989b). A yeast, *Metschnikovia reukaufii* (Ascomycetes), was found to inhabit the nectar of milkweed flowers. It is vectored by flower-visiting insects and, once in the nectar, inhibits germination of pollinia. The nectar is secreted by the stigmatic surface in *Asclepias* and is the natural germination medium for the pollinia, and so it seemed logical to propose an inundative application of the yeast, followed by dispersing it with flower-visiting insects to reduce the fecundity of the plants. Despite the potential utility of manipulating this tri-kingdom system (plant, yeast, and insect) for weed control and the evolutionary implications related to hypotheses for mate selection by plants (Morgan & Schoen, 1997), further research has yet to be done. Nevertheless, the idea did not go unnoticed. Recently, Forcella (2006) has suggested that honey bees, *Apis mellifera* (Apidae), could be used to deliver "microsite-specific gameticides," such as the herbicide glufosinate, to interfere with seed set in weeds.

Control of Plant Diseases

Pollinator biocontrol vector technology has been applied successfully to control gray mold, *Botrytis cinerea* (Moniliaceae), on strawberries, *Fragaria* x *ananassa* (Rosaceae), with the fungal antagonistic, *Clonostachys rosea* (Hypocreales), using honey bees as the pollinator and biocontrol vector (Peng et al., 1992). The levels of control were similar to those obtained from application of a fungicide at the recommended dosages and frequencies. Later, the same technology was applied to raspberries, *Rubus idaeus* (Rosaceae), against the same disease pathogen using honey bees and bumble bees, *Bombus impatiens* (Apidae; Sutton et al., 1996; Yu & Sutton, 1997). The levels of initial success in fruit protection matched or exceeded those achieved with conventional fungicidal sprays. Yu and Sutton (1997) also compared the application of *C. rosea* using a compressed air sprayer versus the pollinator biocontrol vector technology. The incidence of flowers with no inoculum (*C. rosea*) was higher in plots sprayed using the air sprayer (55–57 %) compared with plots treated by bumble bee- (6–9 %) or honey bee- (14–15 %) vectored *C. rosea*. The suppression of gray mold in the flowers was usually greater when *C. rosea* was applied using bees versus the sprayer. Since then, other researchers have applied the fungal mycoparasite *Trichoderma harzianum* (Hypocreaceae) to strawberry flowers using honey bees (Maccagnani et al., 1999) and bumble bees (Kovach et al., 2000), and the fungal antagonist *Ulocladium atrum* (Hypomycetes) to strawberry flowers using honey bees (van der Steen et al., 2006). All reported suppression of gray mold. *Trichoderma* spp. also suppressed *Sclerotinia sclerotiorum* (Sclerotiniaceae), an important crop pathogen. Escande, and co-workers (1994, 2002) successfully used these fungi, vectored by pollinating honey bees, to protect sunflowers, *Helianthus annuus* (Asteraceae), from head rot caused by *S. sclerotiorum*. Svedelius (2000) suppressed the pathogenic plant fungus *Didymella bryoniae* (Ascomycetes) on cucumber, *Cucumis sativus* (Cucurbitaceae), with *T. harzianum* vectored by the bumble bee *Bombus terrestris* (Apidae) in the greenhouse.

A myriad of unknown and potentially useful biocontrol agents can be tried against plant diseases using pollinators as vectors. At about the same time that our research was in progress, Thomson et al. (1992) and Johnson et al. (1993a, 1993b) were experimenting with this technology with honey bees for suppression of infection with the fire blight bacterium, *Erwinia amylovora* (Enterobacteriaceae), using the bacterium *Pseudomonas fluorescens* (Pseudomonadaceae) on apples, *Malus* x *domestica* (Rosaceae), and pears, *Pyrus communis* (Rosaceae). Since then, there has been some resurgence of interest in that system (e.g., Nuclo et al., 1998; Pusey, 2002). Other examples include the fungal plant pathogen mummy berry, *Monilinia vaccinii-corymbosi* (Pezizaceae), which itself is obligately vectored by pollinators to species of blueberry, *Vaccinium* spp. (Ericaceae; Batra, 1983; Woronin, 1888) and can be suppressed by a pollinator-vectored bacterium *Bacillus subtilis* (Bacillaceae; Dedej et al., 2004) and the so-called "killer yeast," *Metschnikovia fructicola* (a newly described ascosporic yeast species), being tested as an inhibitor of gray mold on tender fruit (Kurtzman & Droby, 2001; Karabulut et al., 2003). But as far as we are aware, no one has tried using insect vectors for its dissemination.

Control of Insect Pests

The pollinator biocontrol vector technology has been evaluated successfully against several insect pests on crops. Gross and colleagues (1994) used honey bees to deliver *Heliothis* nuclear polyhedrosis virus (NPHV) to crimson clover, *Trifolium incarnatum* (Fabaceae), to help control *Helicoverpa zea* (Noctuidae), the corn earworm. Although this initiative seems not to have been followed up in the United States, Butt et al. (1998) and, more recently, Carreck et al. (2007) revitalized the idea by applying *Metarhizium anisopliae* (Clavicipitaceae) to the flowers of canola, *Brassica napus* (Brassicaceae), to suppress populations of pestiferous pollen beetles, *Meligethes aeneus* (Nitidulidae), and later, cabbage seed weevils, *Ceutorhynchus assimilis* (Curculionidae). Research by Jyoti and Brewer (1999) demonstrated that honey bees could also be used as effective vectors of the bacterium *Bacillus thuringiensis* var. *kurstaki* (Bt) (Bacillaceae) for control of the banded sunflower moth, *Cochylis hospes* (Tortricidae), in sunflowers, *H. annuus*. The level of control achieved, along with greater pollination efficiency and seed-set, was better than or equivalent to spray applications of Bt.

Our research on using pollinator biocontrol vector technology for insect control started in response to an outbreak of tarnished plant bugs, *Lygus lineolaris* (Miridae), on canola in Alberta, Canada, in 1998 (Cárcamo et al., 2003). We used the entomopathogenic fungus *Beauveria bassiana* (Clavicipitaceae; Bidochka et al., 1993; Gindin et al., 1996) known to cause mortality through disintegration of the insect cuticle and muscle tissues (Bidochka et al., 1993). Realizing that tarnished plant bugs are important pests of numerous crops, including those grown in greenhouses, we expanded our study to include biocontrol of plant bugs on canola as a field crop (figure 5.2) and on sweet pepper, *Capsicum annuum* (Solanaceae), as a greenhouse crop (figure 5.3). Al-mazra'awi et al. (2006a) reported mortalities of tarnished plant bug on canola caged with honey bees using the pollinator biocontrol vector technology at 22–56% versus only 9–22% in the controls. *Beauveria bassiana* conidia were recovered from 100% of the bees sampled in our trials, 67–77% of the flowers, and 70% of the leaves. The mean concentration of inoculum recovered from plant bug samples ranged from 1,411 to 3,803 colony-forming units (CFU) per tarnished plant bug over the 2-year study. For greenhouse peppers, we used bumble bees to vector the *B. bassiana* inoculum, and tarnished plant bug mortality ranged from 34 to 45% versus 9 to 15% in the controls (figure 5.4; Al-mazra'awi,

Figure 5.2 Experimental field set up to evaluate the pollinator biocontrol vector technology using honey bees to vector *Beauveria bassiana* to canola for control of tarnished plant bug. Inset shows honey bee nucleus hive with inoculum dispenser placed inside the cage.

Figure 5.3 Bumble bee hive with inoculum dispenser attached to the front of the hive used in a greenhouse trial set up with sweet peppers.

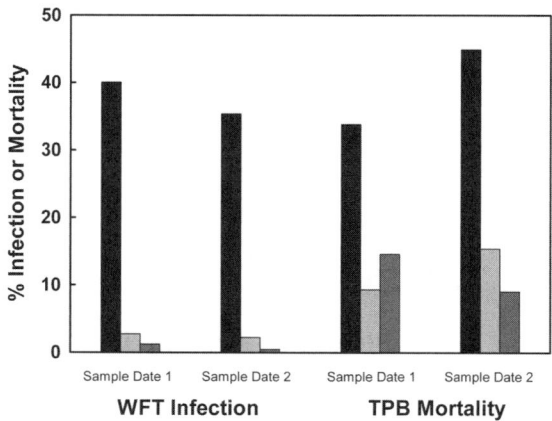

Figure 5.4 Percent infection for western flower thrips (WFT) and percent mortality for tarnished plant bugs (TPB) exposed to bumble bee-vectored *Beauveria bassiana* in greenhouse sweet pepper trials; dark gray bars = treatments in which *B. bassiana* was vectored by bumble bees; light gray bars = treatments with bumble bees only; and medium gray bars = controls.

2004; Al-mazra'awi et al., 2006b). Again, 97–99% of the bees, 90–96% of the flowers, and 87–91% of the leaf samples contained detectable concentrations of *B. bassiana*. In this study, the mean concentration of *B. bassiana* recovered from the tarnished plant bug samples ranged from 587–708 CFUs per tarnished plant bug.

As the project unfolded, the potential for control of western flower thrips, *Frankliniella occidentalis* (Thripidae), a major pest of greenhouse crops, became evident (Al-mazra'awi et al., 2006b). Using bumble bee pollinator biocontrol vector technology, infection rates of western flower thrips ranged from 34 to 40% compared with only 3% in the controls (figure 5.4). Since then, we have further expanded the potential of pollinator biocontrol vector technology to include pests such as greenhouse whitefly, *Trialeurodes vaporariorum* (Aleyrodidae), green peach aphid, *Myzus persicae* (Aphididae), and gray mold in other greenhouse crops, especially tomatoes, *Lycopersicon esculentum* (Solanaceae) (Kapongo et al., 2005; Shipp et al., 2005). Recently, we have demonstrated that pollinator biocontrol vector technology can be used to simultaneously deliver multiple biocontrol agents for insect pest control and disease management (Shipp et al., 2006). Bumble bees have been used to vector an inoculum containing *B. bassiana* and *C. rosea* for tarnished plant bug and whitefly control and for suppression of gray mold on greenhouse sweet pepper and tomato.

Safety for the Vectors and Efficacy Against the Target Pest

When choosing a biocontrol agent and formulation, one must be careful not to unduly endanger the health of the vectors. The potentially useful control agents that suppress seed set in weeds appear to have no adverse effects on the vectors. No matter what sort of agent is considered for use, safety to the vector must be assessed before these biocontrol agents are registered as part of the natural assemblage of microorganisms encountered by pollinators.

The fungal biocontrol agent *T. harzianum* that suppresses the incidence of gray mold has been tested for safety and seems safe to both honey bees and bumble bees (van der Steen et al., 2004). The other plant fungal agents so far used, and potentially useful, are thought to be similarly safe, but they must be tested for vector safety.

However, when considering the use of entomopathogenic agents, one must assume that the risks to the vectors can be measured. Bt has been tested for its effects against adult honey bees (Vandenberg & Shimanuki, 1986) and is regarded as safe; tests on bumble bees have not been made. We have evaluated the risks to pollinators associated with *B. bassiana*. We found that Botanigard 22 WP (Laverlam International Corp., Butte, MT), a commercial formulation of *B. bassiana*, had to be diluted from 2×10^{11} conidia/g of product in the commercial formulation to 6×10^{10} condia/g to achieve minimum mortality of bumble bees (*B. impatiens*) and maximum mortality of the pests (Kapongo et al., 2005; Shipp et al., 2006). At that lower concentration, *B. bassiana* is known to have little effect on honey bees (Vandenberg, 1990; Goettel & Jaronski, 1997). Moreover, it would not be expected to infect at the temperature of the brood chamber (~ 35°C) of the hive. In general, it seems that bumble bees are a little more susceptible than honey bees to developing mycosis from *B. bassiana*, but the risks are small (Al-mazra'awi 2004; Kapongo, unpublished data). *Metarhizium anisopliae* also carries risks to the vectoring bees (Macfarlane, 1976). It is being considered for biological control of *Varroa* mites in honey bee colonies (Kanga et al., 2003; James et al., 2006), so the risks are probably low and similar to those posed by *B. bassiana*. The NPHV used by Gross et al. (1994) is host-specific to Lepidoptera, and as such is probably safe.

Diluents and Formulation

The biocontrol agents used with pollinator biocontrol vector technology are concentrated commercial formulations. They must be diluted to be cost effective and to maximize dispersability and dosage at the target site. Various diluents have been used in preparing inoculum formulations that are placed into the dispensers for use. These diluents can vary greatly in their properties, which can thus affect their utility.

Israel and Boland (1993) found that some carriers, such as talc and especially scented talc, were irritating to honey bees, which can spend as much as 1 minute grooming much of the formulation from their bodies. Other carriers, such as flours, were better accepted, stimulating less than half the amount of time in grooming, and resulted in more efficacious transport of the agent. Spores of *B. bassiana* can be diluted in substrates such as corn flour, corn starch, talc, fungi, or other materials to maximize their dispersability and life span during delivery. Al-mazra'awi (2004) found that honey bees that passed through corn flour acquired more conidia than did bees that passed through wheat flour, durum semolina, corn meal, potato starch, potato flakes, oat flour, and barley flour. As a general rule, the number of conidia carried by the bees increased with decreasing particle size and moisture content of the carrier and with increasing density of *B. bassiana* conidia in the formulation. The time required for a honey bee to pass through the dispenser did not significantly affect the acquisition of conidia. Van der Steen et al. (2006) compared the relative utility of other carrier materials (cellulose, quartz, talc, diatomaceous earth, and clay materials, such as bentonite and kaolin) as to how they affected the adherence of incoculum to honey bees and hence to the petals of strawberry flowers. They suggest that bentonite adheres well to the bodies of bees and is the best in adhering to petals. Polystyrene beads have also been used experimentally (Butt et al., 1998) but may be too costly for use in commercial formulations.

Although full sets of comparisons of various vegetable and mineral diluents remain to be made, we suggest that mineral diluents are likely to be less efficacious because of the irritation that they cause to the bees.

Dispenser Designs

Researchers have used a range of different dispenser designs. Dispensers that direct the pollinators/vectors through the inoculum as they leave the hive and isolate them from the inoculum on entering the hive minimize wastage of the inoculum. This design also reduces the amount of inoculum that enters the hive. At the University of Guelph, we developed a dispenser for dusting honey bees with fungal spores (Peng et al., 1992) that was based on the design of the Nova Scotia Agricultural College pollen dispenser (Townsend et al., 1958; Hatjina, 1988; King & Burrel, 1933). This type of dispenser was used by Butt et al. (1998) and Carreck et al. (2007), among others. Our over-and-under dispenser design forces the bees to pass through the inoculum tray before directing them to depart above the inoculum tray, and bees reenter the hive by passing through the inoculum-free section of the dispenser. The side-by-side dispenser design used by van der Steen et al. (2006) directs the bees to pass through one side to pick up inoculum as they exit but to enter on the other (inoculum-free) side of the dispenser apparatus. Thomson et al. (1992) and Johnson et al. (1993a, 1993b) used the Antles (1953) pollen dispenser, again developed for pollination in pome crops, but it requires bees to both exit and enter the hive through the inoculum. The dispenser developed by Gross et al. (1994) for honey bee delivery of NPHV against *H. zea* on crimson clover is elaborate and commanded its own patent. It was used by Dedej et al. (2004) for dispensing *B. subtilis*. Recently, Bilu et al. (2004) evaluated several types of dispensers for use with honey bees, and they indicate that the University of Guelph dispenser performs well, but that their "Triwaks" has the best overall performance in dusting honey bees with inoculum of *T. harzianum*.

An over-and-under dispenser for bumble bee hives was developed at the University of Guelph (figure 5.5; Yu & Sutton, 1997). In addition, Maccagnani (2005) reported on the comparison of two dispensers, one a side-by-side and the other over-and-under, for the dissemination of fungal antagonists using bumble bees. She reported that the latter is the more effective for delivering fungal biocontrol agents.

Figure 5.5 Dispenser placed in front of a bumble bee hive with the inoculum tray extended.

More studies of dispenser designs for use with both honey bees and bumble bees are needed to address the practical issues of inoculum loading, duration of availability of the inoculum to the bees, rates of inoculum acquisition by the bees, and interference of the dispenser with bee and colony activity, especially through restriction of the hive entrance. The potential also exists for using other insects, including other pollinators, for dispensing biocontrol agents, but few trials have been conducted.

Environmental and Human Safety

Some of the biocontrol agents that have been shown to be useful with this technology are already considered safe from the viewpoint of human health and have been registered with the appropriate agencies in various countries for application to crops. Pollinator biocontrol vector technology is just a different way of applying the biocontrol agent. Nevertheless, registration of microbial control agents is usually specific to the target crop and the method of application. Other biocontrol agents that are under development will require more evaluation for nontarget organism safety (including vector safety) and with respect to residues on food destined to be consumed by human beings or livestock before they can be registered. Thus some environmental and human risk assessment is required before biocontrol agents are to be widely used with this technology. The only registered application of the pollinator biocontrol vector technology that we are aware is Binab (BINAB Bio-Innovation AB, Helsingborg, Sweden). Binab is registered for the application of *Trichoderma* spp. using bumble bees for control of gray mold on strawberries and on greenhouse vegetable crops in several European countries (Biobest B.V., 2006).

Because a number of the biocontrol agents tested for pollinator biocontrol vector technology have rather broad spectra of potential hosts, nontarget risks must be considered. For some agents, such as Bt and NPHV, the risks to nontarget insects are probably small. For others, too few data are available to guide further development. It must be remembered that when considering field crops, the pollinating vectors are unlikely to forage only on the target crop. Thus the same issues that confront conventional spray applications apply: nontarget organisms will be exposed. One may argue that the use of this technology against plant pathogens may be neutral or beneficial to nontarget plants. However, when using pollinator biocontrol vector technology against insect pests, clearly the risks to nontarget, beneficial insects or insects of aesthetic value (Lepidoptera particularly) are real and need investigating. Also, the delivery of microsite-specific gameticides, such as the herbicide glufosinate, to interfere with seed set in weeds (Forcella, 2006) could prove to be problematic to the reproductive output of non-weedy, short-lived plants that flower coincidentally with the target weed.

Discussion and Conclusions

Developing a pollinator vector technology for the management of insect and fungal pests on field crops, such as canola, and on greenhouse crops, such as sweet pepper, provides the benefits of reducing pest populations and pesticide use while improving pollination of the crop. For example, insect pollination of canola improves seed quality and germination rate (Kevan & Eisikowitch, 1990), results in higher seed set and yields (Langridge & Goodman, 1975) and is required for hybrid seed production. Similarly, using bumble bees for the pollination of greenhouse sweet pepper results in increased fruit weight, volume, seed weight, and percentage of extra-large and large fruits and reduced the number of days to harvest (Shipp et al., 1994). Both honey bees and bumble bees can effectively vector biocontrol agents such as *C. rosea*, *T. harzianum*, *B. subtilis*, and *P. fluorescens* against plant pathogens for disease suppression and *B. bassiana*, *M. anisopliae*, Bt, and NPHV to field and greenhouse crops for insect pest control.

Pollinator biocontrol vector technology is a win-win situation because the technology not only reduces pest pressure and pesticide applications but can also improve pollination. Mostly, this technology seems to be safe for the bees, but more laboratory tests, followed by the monitoring of colonies during and after exposure to the biocontrol agents, must be conducted to show no adverse effect on the bees. The development of appropriate formulations and dispensers are key considerations for the success of this technology. The mixture of the dry infective propagules of microbial control agents with diluents and carriers must be made with care to maximize safety and dissemination. Well-formulated agents remain viable for extended time periods in the field and are cost effective. Trials are needed to test each combination of the biocontrol agent and its formulation, the type of pollinator used, the crop to be protected and the pest targeted by the technology, and the sort of dispenser considered to be the most appropriate. Of course, concerns for human and livestock food safety have to be included in the development and registration of any pollinator-biocontrol vector use and need to be coupled with environmental risk assessments with respect to nontarget organisms.

Pollinator biocontrol vector technology is a multidisciplinary pest management approach that incorporates different ecosystem components such as pollinators, microbial control agents and insect pests in the crop production system. It brings the benefits of a new, reduced-risk pest management tool, reduced chemical use, and better pollination of the crop, all of which subsequently result in higher yields and better crop quality.

Acknowledgments

We are grateful to various funding agencies for the research done over the years: the Ontario Pesticides Advisory Board, the Government of Alberta, the Natural Sciences and Engineering Research Council of Canada, the Ontario Beekeepers' Association, the National Research Council of Canada, the Ontario Ministry of Agriculture, Food, and Rural Affairs, Agriculture and Agri-Food Canada, and the Ontario Greenhouse Vegetable Growers' Association. Many people have contributed to the development of the technology, and we acknowledge especially L. Tam, B. Broadbent, S. Khosla, M. Adjaloo, and A. Morse. A. Morse kindly lent his skills to the preparation of figure 5.1.

References

Al-mazra'awi, M. S. (2004). Biological control of tarnished plant bug and western flower thrips by *Beauveria bassiana* vectored by bee pollinators. Unpublished doctoral dissertation, University of Guelph, Ontario, Canada.

Al-mazra'awi, M. S., Shipp, J. L., Broadbent, A. B., & Kevan, P. G. (2006a). Dissemination of *Beauveria bassiana* by honey bees (Hymenoptera: Apidae) for control of tarnished plant bug (Hemiptera: Miridae) on canola. *Environmental Entomology, 35,* 1569–1577.

———. (2006b). Biological control of *Lygus lineolaris* (Hemiptera: Miridae) and *Frankiniella occidentalis* (Thysanoptera: Thripidae) by *Bombus impatiens* (Hymenoptera: Apidae) vectored *Beauveria bassiana* in greenhouse sweet pepper. *Biological Control, 37,* 89–97.

Antles, L. C. (1953). New methods in orchard pollination. *American Bee Journal, 93,* 102–103.

Batra, L. R. (1983). *Monilinia vaccinii-corymbosi* (Sclerotiniaceae): Its biology on blueberry and comparison with related species. *Mycologia, 75,* 131–152.

Bidochka, M. J., Miranpuri, G. S., & Khachatourians, G. G. (1993). Pathogenicity of *Beauveria bassiana* (Balsamo) Vuillemin toward lygus bug (Hem., Miridae). *Journal of Applied Entomology, 115,* 313–317.

Bilu, A., Dag, A., Elad, Y., & Shafir, S. (2004). Honey bee dispersal of biocontrol agents: An evaluation of dispensing devices. *Biocontrol Science and Technology, 14,* 607–617.

Biobest B. V. (2006). Cooperation between Binab® Bio-Innovation Ab and Biobest N. V. for worldwide distribution of the Binab product range [Press release]. Retrieved August 2006 from http://207.5.17.151/biobest/en/nieuws/binab1.htm.

Butt, T. M., Carreck, N. L., Ibrahim, L., & Williams, I. H. (1998). Honey-bee-mediated infection of pollen beetle (*Meligethes aeneus* Fab.) by the insect-pathogenic fungus, *Metarhizium anisopliae*. *Biocontrol Science and Technology, 8,* 533–538.

Cárcamo, H. A., Otani, J., Gavloski, J., Dolinski, M., & Soroka, J. (2003). Abundance of *Lygus* spp. (Heteroptera: Miridae) in canola adjacent to forage and seed alfalfa. *Journal of Entomological Society of British Columbia, 100,* 55–63.

Carreck, N. L., Butt, T. M., Clark, S. J., Ibrahim, L., Isger, E. A., Pell, J. K., et al. (2007). Honey bees can disseminate a microbial control agent to more than one inflorescence pest of oil seed rape. *Biocontrol Science and Technology. 17,* 179–191.

Dedej, S., Delaplane, K. S., & Scherm, H. (2004). Effectiveness of honey bees in delivering the biocontrol agent *Bacillus subtilis* to blueberry flowers to suppress mummy berry disease. *Biological Control, 31,* 422–427.

Eisikowitch, D., Lachance, M. A., Kevan, P. G., Willis, S., & Collins-Thompson, D. L. (1990). The effect of the natural assemblage of microorganisms and selected strains of the yeast *Metschnikovia reukaufii* in controlling the germination of pollen of the common milkweed *Asclepias syriaca*. *Canadian Journal of Botany, 68,* 1163–1165.

Escande, A. R., Laich, F. S., Cuenca, G., Baillez, O., & Pereyra, V. (1994). Dispersíon de inóculo de *Trichoderma* spp. mediante abejas (*Apis mellifera*) para el control de la pudrición de capitulo del girasol (*Sclerotinia sclerotiorum*). *Fitopatología, 29,* 35.

Escande, A. R., Laich, F. S., & Pedraza, M. V. (2002). Field testing of honeybee-dispersed *Trichoderma* spp. to manage sunflower head rot (*Sclerotinia sclerotiorum*). *Plant Pathology, 51,* 346–351.

Forcella, F. (2006, February). Honeybees as novel herbicide delivery systems [Abstract]. In *46th Annual Meeting of the Weed Science Society of America*, New York, New York (90). Retrieved April 2007 from http://ars.usda.gov/SP2UserFiles/Place/36450000/Products-Reprints/2006/1319.pdf.

Gindin G., Barash, I., Raccah, B., Singer, S., Ben-Ze'ev, I. S., & Klein, M. (1996). The potential of some entomopathogenic fungi as biocontrol agents against the onion thrips, *Thrips tabaci* and the western flower thrips, *Frankliniella occidentalis*. *Folia Entomologica Hungarica*, 57(Suppl.), 37–42.

Goettel, M. S., & Jaronski, S. T. (1997). Safety and registration of microbial agents for control of grasshoppers and locust. *Memoirs of the Entomological Society of Canada*, 171, 83–99.

Gross, H. R., Hamm, J. J., & Carpenter, J. E. (1994). Design and application of a hive-mounted device that uses honeybees (Hymenoptera: Apidae) to disseminate *Heliothis* nuclear polyhedrosis virus. *Environmental Entomology*, 23, 492–501.

Hatjina, F. (1998). Hive-entrance fittings as a simple and cost-effective way to increase cross-pollination by honey bees. *Bee World*, 79, 71–80.

Israel, M. S., & Boland, G. J. (1993). Influence of formulation on efficacy of honey bees to transmit biological controls for management of *Sclerotinia* stem rot of canola. *Canadian Journal of Plant Pathology*, 14, 244.

James, R. R., Hayes, G. W., & Leland, J. E. (2006). Field trials on the microbial control of varroa with the fungus *Metarhizium anisopliae*. *American Bee Journal*, 146, 968–972.

Johnson, K. B., Stockwell, V. O., Burgett, D. M., Sugar, D., & Loper, J. E. (1993a). Dispersal of *Erwinia amylovora* and *Pseudomonas fluorescens* by honeybees from hives to apple and pear blossoms. *Phytopathology*, 83, 478–484.

Johnson, K. B., Stockwell, V. O., McLaughlin, R. J., Sugar, D., Loper, J. E., & Roberts, R. G. (1993b). Effect of antagonistic bacteria on establishment of honey bee-dispersed *Erwinia amylovora* in pear blossoms and on fire blight control. *Phytopathology*, 83, 995–1002.

Jyoti, J. L., & Brewer, G. J. (1999). Honey bees (Hymenoptera: Apidae) as vectors of *Bacillus thuringiensis* for control of banded sunflower moth (Lepidoptera: Tortricidae). *Environmental Entomology*, 28, 1172–1176.

Kanga, L. H. B., Jones, W. A., & James, R. R. (2003). Field trials using the fungal pathogen, *Metarhizium anisopliae* (Deuteromycetes: Hyphomycetes) to control the ectoparasitic mite, *Varroa destructor* (Acari: Varroidae) in honey bee, *Apis mellifera* (Hymenoptera: Apidae) colonies. *Journal of Economic Entomology*, 96, 1091–1099.

Kapongo, J. P., Shipp, L., Kevan, P., & Broadbent, B. (2005). Optimal concentration of *Beauveria bassiana* as vectored by bumblebees for pest control on sweet pepper. International Organization for Biological and Integrated Control of Noxious Animals and Plants, West Palaearctic Regional Section 28: 143–146.

Karabulut, O. A., Smilanick, J. L., Mlikota Gabler, F., Mansour, M., & Droby, D. (2003). Near-harvest applications of *Metschnikovia fructicola*, ethanol, and sodium bicarbonate to control postharvest diseases of grape in central California. *Plant Disease*, 87, 1384–1389.

Kevan, P. G., Al-mazra'awi, M. S., Shipp, L., & Broadbent, B. (2004). Bee pollinators vector biological control agents against insect pests: Trials against tarnished plant bug and western flower thrips with the entomopathogenic fungus *Beauveria bassiana* for field canola and greenhouse peppers. In *Memorias 11 Congreso Internacional de Actualización Apícola* (114–117).

Kevan, P. G., Al-mazra'awi, M., Sutton, J. C., Tam, L., Boland, G., Broadbent, B., et al. (2003). Using pollinators to deliver biological control agents against crop pests. In R. A. Downer, J. C. Mueninghoff, & G. C. Volgas (Eds.), *Pesticide formulations and delivery systems: Meeting the challenges of the current crop protection industry* (148–152). West Conshohocken, PA: American Society for Testing and Materials International.

Kevan, P. G., Collins-Thompson, D. L., Eisikowitch, D., & Lachance, M. A. (1989a). Milkweeds, pollinators and yeasts: A system for potential biocontrol of milkweed seed production. *Highlights of Agricultural Research in Ontario, 12,* 21–24.

Kevan, P. G., & Eisikowitch, D. (1990). The effect of insect pollination on canola (*Brassica napus* L. cv. OAC Triton) seed germination. *Euphytica, 45,* 39–41.

Kevan, P. G., Eisikowitch, D., & Rathwell, B. (1989b). The role of nectar in the germination of pollen in *Asclepias syriaca* L. *Botanical Gazette, 150,* 266–270.

Kevan, P. G., Shipp, L., Kapongo, J. P., & Al-mazra'awi, M. S. (2005). Bee pollinators vector biological control agents against insect pests of horticultural plants. In M. Guerra Sanz, A. Roldán Serrano, & A. Mena Granero (Eds.), *First short course on pollination of horticulture plants* (77–95). Almería, Spain: Consejería de Inovación, Ciencia y Impresa, La Mojonera.

Kevan, P. G., Sutton, J. C., Tam. L., Al-mazra'awi, M., Boland, G., Broadbent, B., et al. (2001). Bees as vectors for biological control agents. In *Proceedings of the 7th International Conference on Tropical Bees: Management and diversity and the 5th Asian Apicultural Association conference* (303–306). Cardiff, UK: International Bee Research Association.

King, G. E., & Burrel, A. B. (1933). An improved device to facilitate pollen distribution by bees. *Proceedings of the American Society of Horticultural Science, 29,* 156–159.

Kovach, J., Petzoldt, R., & Harman, G. E. (2000). Use of honeybees and bumble bees to disseminate *Trichoderma harzianum* 1295–22 to strawberries for *Botrytis* control. *Biological Control, 18,* 235–242.

Kurtzman, C. P., & Droby, S. (2001). *Metschnikovia fructicola*, a new ascosporic yeast with potential for biocontrol of postharvest fruit rots. *Systematic and Applied Microbiology, 24,* 395–399.

Langridge, D. F., & Goodman, R. D. (1975). A study on pollination of oilseed rape (*Brassica campestris*). *Australian Journal of Experimental Agriculture and Animal Husbandry, 15,* 285–288.

Maccagnani, B. (2005). Development of devices for the use of bumblebees and mason bees as disseminators of biocontrol agents, and evaluation of their carrying efficiency. Paper presented at the DIARP workshop on the use of pollinators as disseminators of crop protection agents, September 11, 2005, Wageningen, Netherlands.

Maccagnani, B., Mocioni, M., Gullino, M. L., & Ladurner, E. (1999). Application of *Trichoderma hartzianum* by using *Apis mellifera* for the control of grey mold of strawberry: First results. *IOBC Bulletin, 22,*161–164.

Macfarlane, R. P. (1976). Fungi associated with Bombinae (Apidae: Hymenoptera) in North America. *Mycopathologia, 59,* 41–42.

Morgan, M. T., & Schoen, D. J. (1997). Selection on reproductive characters: Floral morphology in *Asclepias syriaca*. *Heredity, 79,* 433–441.

Morse, R. A., & Nowogrodzki, R. (Eds.). (1990). *Honey bee pests, predators, and diseases*. Ithaca, NY: Comstock Press.

Nuclo, R. L., Johnson, K. B., Stockwell, V. O., & Sugar, D. (1998). Secondary colonization of pear blossoms by two bacterial antagonists of the fire blight pathogen. *Plant Disease, 82,* 661–668.

Peng, G., Sutton, J. C., & Kevan, P. G. (1992). Effectiveness of honeybees for applying the biocontrol agent *Gliocladium rosea* to strawberry flowers to suppress *Botrytis cinerea*. *Canadian Journal of Plant Pathology, 14*, 117–129.

Pusey, P. L. (2002). Biological control agents for fire blight of apple compared under conditions limiting natural dispersal. *Plant Disease, 86*, 639–644.

Shaw, D. E. (1999). Bees and fungi, with special reference to certain plant pathogens. *Australasian Plant Pathology, 28*, 269–282.

Shipp, J. L., Whitfield, G. H., & Papadopoulos, A. P. (1994). Effectiveness of the bumble bee, *Bombus impatiens* Cr. (Hymenoptera: Apidae), as a pollinator for greenhouse sweet pepper. *Scientia Horticulturae, 57*, 29–39.

Shipp, L., Kapongo, J. P., Kevan, P., Sutton, J., & Broadbent, B. (2006). Bumble bees: An effective delivery system for microbial control agents for arthropod pest and disease management. International Organization for Biological and Integrated Control of Noxious Animals and Plants, West Palaearctic Regional Section 29, 47–51.

Sutton, J. C., Li, D. W., Peng, G., Yu, H., Zhang, P., & Valdebenito-Sanhueza, R. M. (1996). *Gliocladium rosea*: A versatile adversary of *Botrytis cinerea* in crops. *Plant Disease, 81*, 316–326.

Svedelius, G. (2000). Humlor som barare av biologisk control av svampsjukdomen svartprickrota gurkfrukter. *Vaxtskyddsnotiser, 64*, 48–50.

Thomson, S. V., Hansen, D. R., Flint, K. M., & Vandenberg, J. D. (1992). Dissemination of bacteria antagonistic to *Erwinia amylovora* by honey bees. *Plant Disease, 76*, 1052–1056.

Townsend, G. F., Riddle, R. T., & Smith, M. V. (1958). The use of pollen inserts for tree fruit pollination. *Canadian Journal of Plant Sciences, 38*, 39–44.

van der Steen, J. J. M., Donders, J., & Blacquière, J. (2006). The use of honeybees as disseminators of *Ulocladium atrum* against grey mold in strawberries. Retrieved June 2006 from http://documents.plant.wur.nl/ppo/bijen/antagonisten.pdf.

van der Steen, J. J. M., Langarak, C. J., van Togeren, C. A. M., & Dik, A. J. (2004). Aspects of the use of honeybees and bumblebees as vectors of antagonistic micro-organisms in plant disease control. *Proceedings of the Netherlands Entomological Society, 15*, 41–46.

Vandenberg, J. D. (1990). Safety of four entomopathogenic fungi for caged adult honeybees (Hymenoptera: Apidae). *Journal of Economic Entomology, 83*, 755–759.

Vandenberg, J. D., & Shimanuki, H. (1986). Two commercial preparations of the beta exotoxin of *Bacillus thuringiensis* influence the mortality of caged adult honeybees *Apis mellifera* (Hymenoptera: Apidae). *Environmental Entomology, 15*, 166–169.

Wodehouse, R. P. (1959). *Pollen grains: Their structure, identification, and significance in science and medicine*. New York: Hafner.

Woronin, M. (1888). Über die Sclerotienkrankheit der Vaccinieen-Beeren. *Mémoires de l' Academie Imperiale des Sciences de St.-Pétersbourg (VIIe Série), 36*, 1–49.

Yu, H., & Sutton, J. C. (1997). Effectiveness of bumblebees and honeybees for delivering inoculum of *Gliocladium roseum* to raspberry flowers to control *Botrytis cinerea*. *Biological Control, 10*, 113–122.

Part 2

Managing Solitary Bees

6 Life Cycle Ecophysiology of *Osmia* Mason Bees Used as Crop Pollinators

Jordi Bosch, Fabio Sgolastra, and William P. Kemp

Introduction

Agricultural ecosystems planted with insect-pollinated plants create a situation of high pollinator demand over a short blooming period. This is especially true in areas of intensive monoculture, in which alternative flowering species are frequently unavailable to pollinators prior to and/or following petal fall of the main crop. Other environmental conditions in these areas, such as the destruction of natural nesting sites and the use of pesticides and herbicides, also contribute to the demise of wild pollinator populations. Where measures to protect and enhance these populations (see chapter 2, this volume) are difficult to implement, it becomes necessary to introduce populations of managed pollinators. This need has stimulated the search for pollinator species suitable to various crops and agronomic environments (open fields, greenhouses, screened enclosures). Some of these efforts have resulted in the development of viable pollinator management systems.

"Pollinator management" describes a range of situations and practices. In its simplest form, pollinator management consists of a set of measures to provide adequate and safe nesting sites and foraging conditions (e.g., nesting substrates, continuous bloom, pesticide-free environment) to enhance resident pollinator populations. This approach has been successful in managing populations of two ground-nesting halictid bees, the alkali bee (*Nomia melanderi*) in the United States and the alfalfa gray-haired bee (*Rhophitoides canus*) in eastern Europe (Bohart, 1958; Stephen, 1960; Johansen, Mayer, Stanford, & Kious, 1982; Ptacek, 1989). In a second level of intervention and complexity, pollinator management applies to those situations in which populations of pollinators are reared on the target crop and then stored under more or less controlled conditions to reduce mortality and/or manipulate pollinator phenology (e.g., developmental rates, number of generations, emergence time). This is accomplished by providing appropriate temperatures for development and wintering, supplementing food resources, and reducing parasitism and predation. Honey bees (*Apis mellifera*, Apidae; Crane, 1991; Free, 1993), alfalfa leafcutting bees (*Megachile rotundata*, Megachilidae; Bohart, 1962; Stephen, 1962; Hobbs, 1973; Richards, 1984; see also chapter 8, this volume), and mason bees (*Osmia* spp., Megachilidae; Yamada, Oyama, Sekita, Shirasaki, & Tsugawa, 1971; Maeta & Kitamura, 1974; Bosch & Kemp, 2001; Torchio, 2003; Krunic & Stanisavljevic, 2006) are usually managed at this level. A third level of management is applied to species such as bumble bees (*Bombus* spp., Apidae) that are reared under highly artificial laboratory conditions and then taken to the target crop (van Heemert, de Ruijter, van den Eijnde, & van der Steen, 1990; Asada & Ono, 2002; Velthuis & van Doorn, 2006).

Despite the economic value of bee crop pollination (Southwick & Southwick, 1992), and despite the far-from-negligible contribution of a diversity of wild bees to crop pollination (see chapter 2, this volume), only a handful of species have been developed into manageable crop pollinators. Often, studies on potential crop pollinators have emphasized pollinating effectiveness. However, the pollinating contribution of a highly effective pollinator is likely to remain anecdotal unless a management system (at any of the three levels outlined earlier) and a reliable population supply are secured. Our goal in this chapter is to draw attention to the importance of developing adequate rearing methods to establish new pollinator species and improve performance of existing ones. In our opinion, many highly efficient pollinators have remained "potential pollinators" because of a lack of basic studies on their developmental biology and the establishment of appropriate rearing methods. A parallel can be drawn with insects established as biocontrol agents of agricultural pests. Extensive use of these insects has been possible thanks to a solid understanding of the ecophysiology of their life cycles (e.g., Wajnberg & Hassan, 1994; Hodek & Honěk, 1996).

In this chapter, we first review our current knowledge on the developmental biology of *Osmia* crop pollinators. Then we discuss how this knowledge has contributed and continues to contribute to their establishment as manageable pollinators. We make extensive reference to current practices used to manage populations of these species and discuss how some of these practices may affect bee development, vigor, and survival. At the end of the chapter we establish a comparison with *Megachile rotundata*, a species widely used to pollinate alfalfa (see chapter 7, this volume). Although *Osmia* and *Megachile* are phylogenetically related and share many behavioral traits, their life cycles differ significantly. Thus the comparison between the two genera is useful to emphasize developmental physiology as a key factor in establishing appropriate rearing methods and management systems.

Osmia *Crop Pollinators*

The genus *Osmia* comprises more than 300 species, mostly in the Holarctic (Michener, 2000). The majority of these species nest in preestablished cavities in which females build series of cells separated by partitions made of mud or masticated leaf material.

Osmia in general, and species in the subgenus *O. (Osmia)* in particular, fly very early in the year. In part for this reason, several *O. (Osmia)* species have been developed in different parts of the world to pollinate spring-blooming crops.

Osmia cornifrons was developed as an orchard pollinator in Japan in the 1960s and is now being used on more than 70% of the apple acreage in that country (Yamada et al., 1971; Maeta & Kitamura, 1974; Maeta, 1990). More recently, this species has been used in China and Korea (Xu, Yang, & Kwon, 1995). In the late 1970s and early 1980s, *Osmia cornifrons* populations from Japan were introduced into the eastern United States, where the species is now becoming established as an orchard pollinator (Batra, 1979, 1998). *Osmia cornifrons* has also been tested on blueberries, several greenhouse crops (including strawberries, melons, and watermelons), and caged legume and mustard crops (Maeta, 1974; Maeta, Okamura, & Ueda, 1990; Abel, Wilson, & Luhman, 2003; Maeta, Nakanishi, Fujii, & Kitamura, 2006). A North American sister species, *O. lignaria*, has been developed in the United States for orchard pollination, and its use in commercial orchards is steadily increasing (Torchio, 1976, 1985, 2003; Bosch & Kemp, 2001, 2002; Bosch, Kemp, & Trostle, 2006). Additionally, *O. lignaria* has been tested on blueberries in Canada and on caged Brassicaceae in the United States (Dogterom, 1999; Abel et al., 2003). In Europe, a third species, *O. cornuta*, has been developed as an orchard pollinator (Asensio, 1984; Bosch, 1994a; Vicens & Bosch, 2000; Maccagnani, Ladurner, Santi, & Burgio, 2003; Ladurner, Recla, Wolf, Zelger, & Burgio, 2004; Monzón, Bosch, & Retana, 2004; Krunic & Stanisavljevic, 2006). *Osmia cornuta* also has been successfully used in enclosures to produce hybrid seed of Brassicaceae (Ladurner, Santi, Maccagnani, & Maini, 2002).

Other *Osmia* species have been tested to various levels, often with successful results. These include *O. (Osmia) rufa* on fruit trees and caged mustard crops for hybrid seed production (Holm, 1973; Roth, 1990; van der Steen & de Ruijter, 1991; O'Toole, 2002; Steffan-Dewenter, 2003), *O. (Osmia) ribifloris* and *O. (Melanosmia) attriventris* on blueberries (Drummond & Stubbs 1997; Torchio, 1990), *O. (Osmia) excavata* on apples (Wei, Wang, Smirle, & Xu, 2002), *O. (Helicosmia) caerulescens* and *O. (Melanosmia) sanrafaelae* on alfalfa (Taséi, 1972; Parker, 1981, 1989), and *O. (Melanosmia) aglaia* on raspberries and blackberries (Cane, 2005).

Osmia Life Cycle

The aforementioned *O. (Osmia)* and *O. (Melanosmia)* species fly very early in the year, are univoltine, and spend the winter as cocooned adults. Individuals wintered as prepupae or pupae are not viable (Bosch, 1994a; Bosch & Kemp, 2000; Bosch et al., 2006). However, *O. ribifloris* have been successfully wintered as prepupae under artificial conditions (J. H. Cane, personal communication). *Osmia (Helicosmia) caerulescens* also winters in the adult stage but has been found to be bivoltine or partially bivoltine throughout its distribution range (Krombein, 1967; Taséi, 1972; Vicens, Bosch, & Blas, 1993; Westrich, 1989). Wintering in the adult stage appears to be a derived trait within the Megachilidae.

Most members of this family, including the ancestral Lithurginae, winter as prepupae and fly later in the year (Bosch, Maeta, & Rust, 2001). Spring-flying species of other bee genera (e.g., *Anthophora* and *Colletes*) also winter as adults, in contrast to summer-flying species of the same genera that winter as prepupae (Westrich, 1989). Thus adult wintering appears to be associated with early-flying periods. Many species in *Helicosmia* and other *Osmia* subgenera have been found to be parsivoltine (Bosch et al., 2001; Torchio & Tepedino, 1982). That is, some individuals within a population are univoltine, whereas others are semivoltine (take 2 years to develop). In parsivoltine *Osmia* species, univoltine individuals winter as adults, and semivoltine individuals spend the first winter as prepupae and the second as adults. Semivoltinism is highly dependent on temperature, and its incidence increases with altitude (Bosch, Kemp, & Sgolastra, unpublished).

The focus of this chapter is on *O. lignaria*, *O. cornifrons*, and *O. cornuta*, the three species whose developmental biology is best known and for which management methods have been further developed. To our knowledge, all populations of these three species studied to date are strictly univoltine. Their life cycle can be divided into six periods (see figure 6.1):

1. *Emergence and mating.* Wintered adults are exposed to warm temperatures (spring incubation), emerge from the cocoon, leave their natal nest, and mate.
2. *Prenesting.* During a brief period, females mature their ovaries prior to nesting.
3. *Nesting.* Females build nests and lay eggs in provisioned cells.
4. *Development.* Eggs develop into adults through five larval instars, a prepupal dormant stage, and a pupal stage.
5. *Prewintering.* Newly eclosed adults inside their cocoons remain exposed to warm temperatures at the end of the summer and beginning of autumn.
6. *Wintering.* Cocooned adults are exposed to cold temperatures.

Within a species, there are important phenological differences among populations. In warmer areas, adults fly and nest as early as February, as compared with April–May in colder areas. We use the term "early-flying" to describe populations that nest in February–March and "late-flying" to describe populations that nest in April–May.

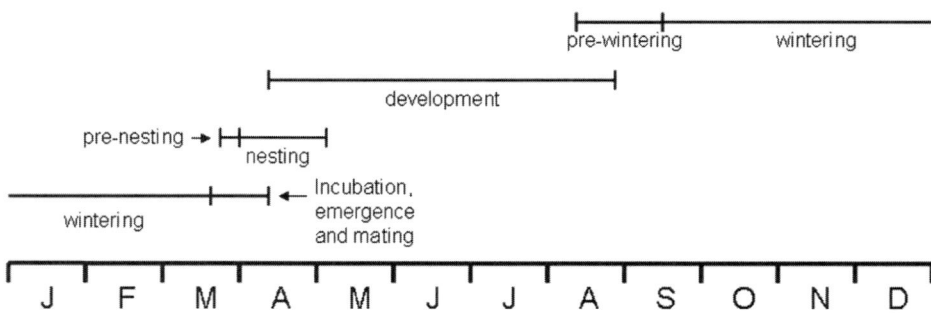

Figure 6.1 Life cycle and phenology of a univoltine *Osmia* population. Letters refer to months of the Roman calendar.

Emergence and Mating

Although male and female emergence periods overlap, males emerge, on average, 2–4 days ahead of females. Emerged males can be seen flying around nesting areas and congregating at the entrance of nests that newly emerged females are about to leave. This behavior occurs in response to a short-lived sex pheromone released by newly emerged females (Rösner, 1994). Mating often takes place as soon as females leave the natal nest. Males stacked on top of newly emerged females are a common sight at nest aggregations on sunny days during the emergence period. However, some mating also occurs on nearby flowers.

Timing of emergence is highly dependent on wintering duration and spring incubation temperatures. With few exceptions, individuals not wintered or wintered for very short periods do not emerge (see the later section on wintering), irrespective of incubation conditions. On the other hand, in populations exposed to long wintering periods or to mild wintering temperatures, some individuals (mostly males) emerge at wintering temperatures (i.e., without incubation; Bosch & Blas, 1994; Monzón, 1998; Bosch & Kemp 2003, 2004). In both *O. lignaria* and *O. cornifrons*, mean emergence time decreases with increasing incubation temperature (from 15 to 30°C; Bosch & Kemp, 2001; Maeta et al., 2006). However, survival in *O. lignaria* is slightly lower at 15 and 30°C than at 20 or 25°C. Temperature thresholds for emergence are much lower for males than for females. In late-flying *O. lignaria* populations incubated at ambient (outdoor) temperatures, male emergence proceeds at temperatures of 12–15°C, but female emergence is very slow unless temperatures reach 20°C (Bosch & Kemp, 2001). These differences between males and females are important when developing incubation protocols. Because female bees perform most of the pollination service, any models used to predict emergence and facilitate the timing of peak bee activity to peak orchard bloom should be based on female temperature requirements.

Prenesting

Immediately following mating, females are rarely seen around their natal nesting site, only to appear back at the nesting site after a few days. *Osmia* oocytes are not fully grown at emergence, and females complete ovary maturation during the prenesting period (Monzón, 1998; Sgolastra, 2007). Duration of the prenesting period is usually 2–5 days, but it may be longer under bad weather conditions (Maeta, 1978; Bosch & Kemp, 2001; Bosch & Vicens, 2006). Prenesting periods need to be considered when deciding incubation schedules for managed populations.

During the prenesting period, females take shelter at night in crevices on walls or on the bark of trees, sometimes in clusters containing large numbers of males. Over time, increasing numbers of females are found hovering around nesting sites and entering prospective nest cavities. At artificial nesting sites (with many nesting cavities available), the cavity-selection process may continue for several hours, during which time each female repeatedly investigates several cavities in succession. This behavior is punctuated by short nectar-collecting trips. Eventually, a female limits her visits to a particular cavity. A zig-zagging flight with increasingly wide loops in front of the nesting site usually signals the selection of a nesting cavity and precedes the initiation of nesting activities.

Nesting

Having selected a cavity in which to nest, females begin collecting nesting material (mud for *O. lignaria*, *O. cornifrons*, and *O. cornuta*) to build a basal partition and pollen and nectar to start the first provision. Detailed accounts of nesting behavior can be found in Maeta (1978), Torchio (1989), Bosch (1994b), and Bosch & Kemp (2001). Nests typically consist of a series of cells separated by mud partitions. Each cell contains a loaf of pollen mixed with nectar, on top of which an egg is laid. The entrance of completed nests is sealed with a thick mud plug. Individual females are active for 20–25 days (including the prenesting period) and build 0.5 to 1.5 cells per day under field conditions (Maeta, 1978; Torchio, 1989; Bosch & Vicens, 2005, 2006; Bosch, 2008). In orchard environments, each female may build a total of 8–12 cells, of which 2.5 to 5 contain female progeny (see Bosch & Kemp, 2002, and references therein; Maeta, 1978; Bosch & Vicens, 2006). Production of female progeny declines throughout the nesting period, and aged females produce almost exclusively male progeny (Tepedino & Torchio, 1982; Sugiura & Maeta, 1989; Bosch & Vicens, 2005).

Egg and Larval Development

Under field conditions, *Osmia* eggs take more than a week to hatch (Bosch & Kemp, 2000). Larval development proceeds through five instars (Torchio, 1989) and takes approximately a month under field conditions (Bosch & Kemp, 2000). Temperature thresholds for egg and larval development have been estimated at 10–14°C and 7–14°C, respectively (Maeta, 1978; Maeta et al., 2006; Sgolastra, 2007). In *O. lignaria* and *O. cornifrons*, egg and larval developmental rates increase with increasing temperatures from 18 to 26°C, and then stabilize at 29–30°C (Bosch & Kemp, 2000; Maeta et al., 2006). On consuming the pollen-nectar provision and completing defecation, the fifth instar spins a thick multilayered cocoon with secretions originating from the salivary glands (Torchio, 1989). Cocoon spinning takes 4–8 days at 22–26°C (Bosch & Kemp, 2000; Maeta et al., 2006). When exposed to the same temperatures, early-flying populations develop slightly slower than late-flying populations. For example, at 26°C egg and larval developmental took 26 days in an *O. lignaria* March-flying population compared with 20 days in an April-flying population (Bosch & Kemp, 2000; Sgolastra, 2007).

Prepupal Dormancy

Cocooned larvae (prepupae) enter a dormant stage (Torchio, 1989) more or less in synchrony with early summer (Bosch & Kemp, 2000). CO_2 production rates (measured at 22°C) drop from ~ 0.90 ml/g·h in the feeding larva to ~ 0.14 immediately after cocoon completion, and then to 0.04–0.09 within 7 days of cocoon completion (Kemp, Bosch, & Dennis, 2004; Bosch & Kemp, unpublished). As respiration rates decline, the prepupa becomes progressively flaccid in coincidence with the lowest respiration point. Then, toward the end of the prepupal period, the prepupa becomes turgid again as respiration rates increase to levels of 0.14–0.18 ml/g·h. Eventually, as the prepupa approaches pupation, the segmentation that defines the head, thorax, and abdomen becomes apparent (Kemp et al., 2004). If not exposed to appropriate temperatures (see later in the chapter), *Osmia* may remain in the prepupal stage for months or even years without dying (Bosch 1994a; Bosch & Kemp, 2000; Maeta et al., 2006; Sgolastra, 2007). Body weight loss during the prepupal period is only ~ 0.08 mg/day, compared with 0.3–0.7 during the pupal period (Bosch & Vicens, 2002; Kemp et al., 2004). These results, together with temperature requirements for prepupal development (see later in the chapter), indicate that prepupal dormancy in *Osmia* is diapause-mediated (*sensu* Tauber, Tauber, & Masaki, 1986). In comparison, the prepupal respiration rates of wintering *M. rotundata* are maintained low through the summer and winter (Kemp et al., 2004).

Under natural conditions, prepupal dormancy in *Osmia* lasts 1–3 months, depending on the geographical area (Bosch & Kemp, 2000; Bosch, Kemp, & Peterson, 2000; Sgolastra, 2007). Temperatures above a certain threshold are required for prepupal diapause completion. Prepupae of *O. cornifrons* may be kept at 10°C for up to 430 days without dying (Maeta et al., 2006). If subsequently exposed to temperatures of 22–26°C, these individuals develop into apparently healthy adults. In *O. lignaria* reared at 18 or 20°C, some individuals fail to terminate diapause and may remain in the prepupal stage for over a year (Bosch & Kemp, 2000; Sgolastra, 2007). If these individuals are subsequently exposed to a 22°C temperature, some develop into adults but do not survive a wintering period. When individuals failing to pupate at 18 or 20°C are wintered as prepupae and then incubated, a few of these prepupae resume development, but they rarely develop into healthy adults; if they do, they emerge out of synchrony with their natural life cycle. Similar results have been obtained in *O. cornuta* reared at 22°C (Bosch, 1994a).

Maximum prepupal developmental rates are attained at intermediate (22–26°C) temperatures (Kemp & Bosch, 2005; Maeta et al., 2006; Sgolastra, 2007). At warmer temperatures (26–32°C) prepupal development is slower. In nature, however, *Osmia* nests are exposed to fluctuating temperature regimes, and the thermoperiod has a strong effect on prepupal diapause completion. In *O. cornuta*, 15% of individuals exposed to 22°C failed to pupate, as compared with 100% pupation when exposed to a daily cycle of 12 h at 17°C and 12h at 27°C (mean = 22°C; Bosch, unpublished). Prepupal period in *O. lignaria* lasts 29 days at 22°C constant, as compared with 17 days when exposed to 8h at 14°C and 16h at 27°C (mean = 22°C; Bosch & Kemp, 2000). In general, populations kept under fluctuating temperature regimes have lower developmental mortality.

Populations from different areas differ in their thresholds for prepupal diapause completion. The percentage of prepupae failing to pupate at 25°C was 12% in a February-flying *O. cornuta* population compared with 2% in a March-flying population exposed to the same conditions (Bosch, 1994a). Similar results were obtained in *O. lignaria*. At 20°C,

56% of the individuals in a March-flying population failed to pupate compared with 23% in an April-flying population held at 18°C (Sgolastra, 2007). Above these thresholds, early-flying populations have longer prepupal developmental periods than late-flying populations exposed to the same temperatures. At 22°C, progeny from a February-flying *O. cornuta* population took 95 days to develop from cocoon completion to pupation compared with 75 days in an April-flying population (Bosch unpublished). In *O. lignaria*, prepual period at 26°C was 52 days in March-flying populations compared with 29 days in April-flying populations (Bosch & Kemp, 2000; Sgolastra, 2007). These differences persist under natural conditions. March-flying populations kept outdoors took more than 2 months to pupate compared with approximately 1 month in April-flying populations (Bosch et al., 2000; Sgolastra, 2007). When bees from a late-flying population from Utah were forced-reared and released in a February-blooming almond orchard in California, their progeny expressed short prepupal periods when reared under both California conditions and laboratory conditions (Bosch et al., 2000).

These results suggest a significant genetic component for diapause development in *Osmia* and indicate a likely adaptation to local temperature conditions (Ayres & Scriber, 1994). Duration of the prepupal period could also be influenced by a phenological maternal effect (Mousseau & Dingle, 1991). For example, diapause development could be dependent on photoperiod conditions experienced by the maternal generation, as suggested for *M. rotundata* (Parker & Tepedino, 1982; Kemp & Bosch, 2001). In an unpublished study, an April-flying *O. lignaria* population was divided into two groups. The first group was released on almonds in February, 2 months ahead of its normal flying time, and the second group was released on cherries in April, the usual flying time for this population (Bosch & Kemp, unpublished). The progeny of both groups were reared under the same conditions. In agreement with a possible maternal effect, prepupal duration was significantly longer in the first group, but only by 2–6 days, a very short period compared with the 2-month difference in flying time between the two parental groups.

Pupation

Newly formed pupae take 1 to 1.5 months to acquire adult pigmentation (Bosch & Kemp, 2000; Sgolastra, 2007). Respiration rates (measured at 22°C) throughout the pupal period are 0.14–0.18 CO_2 ml/g·h, and body weight loss rate is 0.3–0.7 mg/day (Bosch & Vicens, 2002; Kemp et al., 2004; Bosch & Kemp, unpublished). Temperature thresholds for pupal development have been estimated at 10–12°C (Maeta et al., 2006; Sgolastra, 2007). In contrast to prepupal development, and within the temperature range tested (18–30°C), pupal developmental rates increase with temperature (Bosch & Kemp, 2000; Maeta et al., 2006; Sgolastra, 2007). Also in contrast to prepupal development, thermoperiod does not have a profound effect on pupal developmental rates. In *O. lignaria*, the pupal stage lasts 32 days at 22°C, as compared with 28 days when exposed to a cycle of 8h at 14°C and 16h at 27°C (mean = 22°C; Bosch & Kemp, 2000). In *O. cornuta* the pupal stage lasted 36 days both at 22°C and when exposed to 12h at 17°C and 12h at 27°C (mean = 22°C; Bosch, unpublished). Differences in pupal period between populations of different geographic origin are not as pronounced as differences in prepupal period duration. When exposed to 22°C, *O. cornuta* early- and late-flying populations expressed similar pupal duration (36 and 35 days, respectively; Bosch, unpublished). In *O. lignaria* at 26°C, pupation of an early-flying population lasted 30 days compared with 24 days in a late-flying population (Bosch & Kemp, 2000; Sgolastra, 2007).

Prewintering

At the end of the pupal period, the pupa sheds its cuticle and the callow adult ecloses and stretches its wing pads. Adult eclosion takes place at the end of summer or beginning of autumn. Early-flying *O. lignaria* populations from central California reach adulthood in September compared with July–August in late-flying populations from northern Utah (Bosch & Kemp, 2000; Bosch et al., 2000; Sgolastra, 2007). During prewintering the cocooned adult is still exposed to temperatures warm enough for development. This period allows slowly developing individuals within a population to reach adulthood and be ready for wintering when temperatures start to decline. As mentioned, differences in developmental rate between populations of different geographical origin exposed to the same temperatures are greater for the prepupal stage than for the larval or pupal stages. At the same time, most variation in timing of adult eclosion among individuals within a population is explained by the duration of the prepupal stage (Sgolastra, 2007). Thus, duration of the prepupal diapause is the main mechanism through which *Osmia* populations from different geographical areas synchronize adult eclosion with local temperature declines in the autumn (Kemp & Bosch, 2005; Sgolastra, 2007).

Within 5 days of adult eclosion, CO_2 production rate starts a precipitous decline, from 0.14–0.18 to ~ 0.10 ml/g·h in 14 days (Kemp et al., 2004; Bosch, Kemp, & Sgolastra, unpublished). Together with the response of wintering adults to incubation (discussed later), these results indicate that, as with prepupal (summer) dormancy, adult (winter) dormancy is diapause-mediated in *Osmia*. The decline in respiration activity shortly after adult eclosion occurs under natural conditions, but also in individuals kept at constant 22°C. Thus, a temperature cue is not required for diapause induction. Respiration rates (measured at 22°C) in diapausing prewintering adults do not increase until exposure to cold temperatures (Bosch, Kemp, & Sgolastra, unpublished). Thus, from a physiological perspective prewintering may be defined as the period between adult eclosion and the increase in respiration rates prompted by the onset of winter temperatures.

Prewintering duration in *Osmia* has important consequences for diapause development, winter survival, and vigor at emergence in spring. Maximum survival is obtained in individuals prewintered for short to intermediate (15–30 days) periods (Monzón, 1998; Bosch et al., 2000; Bosch & Kemp, 2004; Sgolastra, 2007; Bosch, et al., unpublished). Individuals prewintered for longer periods (45, 60, or 80 days) maintain low respiration rates (~ 0.1 ml/g·h) but rapidly lose weight (0.2–0.4 mg/day) and express signs of extensive fat body depletion. Their lipid levels drop significantly. These individuals are less likely to survive the winter and are less vigorous at emergence than individuals prewintered for 15–30 days. Individuals prewintered for very short periods (3–5 days) do not reach minimum respiration levels of ~ 0.08 ml/g·h. Nonetheless, these individuals respond to cold-temperature exposure by increasing their respiration (measured at 22°C) in a similar way to individuals prewintered for 15–30 days (Sgolastra, 2007). Individuals prewintered for very short periods show no signs of fat body depletion, maintain high lipid levels, and are more vigorous at emergence than individuals prewintered for 15–30 days. However, winter survival in individuals prewintered for very short periods is usually 5–10% lower than in individuals prewintered for 15–30 days (Monzón, 1998; Bosch et al., unpublished). Individuals not prewintered at all (cooled on the day of adult eclosion) do not survive the winter (Maeta et al., 2006). Biochemical adjustments in preparation for wintering in insects are often initiated before the onset of cold temperatures (Denlinger, 1991). In *O. cornifrons*, exposure to warm temperatures in autumn appears to be necessary to build up levels of polysaccharides (Maeta et al., 2006), which would explain the low survival in individuals wintered as teneral adults.

Populations from colder areas emerge and nest later in the year, and their progeny are exposed to a shorter summer. Thus their progeny need to reach adulthood earlier in the year within a shorter time frame. As mentioned, faster development from egg to adult in these populations is mainly achieved through a shorter prepupal diapause (figure 6.2). Within a geographical area, timing of adult eclosion in relation to the onset of wintering temperatures is subject to significant temporal variation (Kemp & Bosch, 2005; Sgolastra, 2007; Bosch et al., unpublished;). First, temperatures in late summer and early autumn vary significantly from year to year. Second, the timing of adult eclosion also varies from year to year, both in absolute terms and in relation to the onset of wintering temperatures. Finally, eclosion time within a population may span more than 1 month, and thus there are important differences among individuals in adult eclosion date. As mentioned, individual variability in adult eclosion date is mostly explained by variability in individual duration of the prepupal diapause. The heritability of prepupal diapause developmental rate in *Osmia* has not been studied. However, a genetic component for diapause development has been found in various insect species (Tauber et al., 1986; McWatters & Saunders, 1996; Bradshaw, et al., 1997; Feder et al., 1997; Gomi, 1997), and evidence suggests that this genetic basis also occurs in *Osmia* (see the section on prepupal dormancy). Under this scenario, and given the consequences of individual prewintering duration on winter survival, selection would favor those phenotypes developing early enough to reach adulthood before the onset of winter but late enough to avoid exposure to an excessively long prewintering period.

Wintering

Respiration rates (measured at 22°C) climb steadily throughout the wintering period from a 0.08 ml/g·h prewintering low to 0.4 ml/g·h just prior to emergence (Kemp et al., 2004; Sgolastra, 2007). Weight loss rate during wintering (0.05–0.09 mg/day) is low

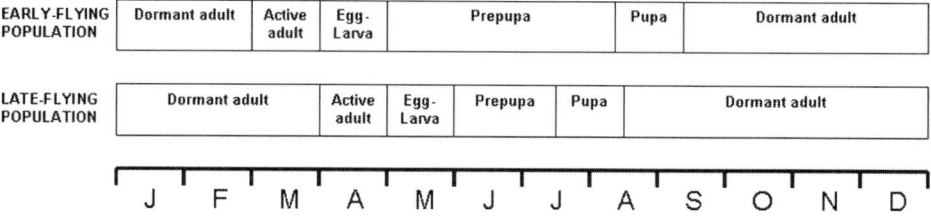

Figure 6.2 Approximate developmental phenology in early- and late-flying *Osmia*. Letters refer to months of the Roman calendar.

compared with prewintering, as is fat body depletion (Bosch & Vicens, 2002; Bosch & Kemp, 2003, 2004; Kemp et al. 2004). Individuals lose a higher percentage (~10%) of their body weight during 1 month of prewintering than during 5 months of wintering (~7%).

Exposure to cold temperatures is necessary for adult diapause development in *Osmia*. As mentioned in the previous section, *Osmia* adults not exposed to wintering temperatures maintain low respiration rates, lose weight rapidly, deplete their fat bodies, and eventually die. Survival is also low (0–40%) in individuals wintered for very short periods (30 days; Bosch & Kemp, 2003, 2004; Maeta et al., 2006). After 30 days of wintering, respiration rates (measured at 22°C) are ~0.2 ml/g.h, compared to ~0.4 in fully wintered (~200 days) individuals (Sgolastra, 2007). When individuals wintered for only 30 days are incubated at 20°C, their respiration rates rapidly decrease and go back to the minimum levels attained during prewintering (Kemp et al., 2004; Sgolastra, 2007). These individuals maintain low respiration rates but rapidly lose weight as incubation proceeds. Adults taken out of the cocoon and exposed to room temperature in midwinter are able to walk but do not fly or attempt to feed when offered a flower (Bosch & Kemp, 2003).

Preemergence time (time required to emerge following spring incubation) of wintered *Osmia* adults is inversely proportional to wintering duration (Bosch & Blas, 1994; Bosch & Kemp, 2003, 2004; Maeta et al., 2006). For example, *O. lignaria* males wintered at 4°C for 90 days take 14 days to emerge when incubated at 20°C compared with 3 days when wintered for 150 days and 1 day when wintered for 210 days (Bosch & Kemp, 2003). These results match the steady increase in respiration rates (measured at 22°C) observed throughout wintering (Kemp et al., 2004; Sgolastra, 2007) and corroborate the need for a cold period to complete diapause in *Osmia*.

When exposed to sufficiently long wintering periods, diapause development is faster at higher versus lower wintering temperatures (Bosch & Kemp, 2003, 2004). For example, *O. lignaria* males wintered at 4°C for 210 days take 1 day to emerge following incubation at 20°C in comparison with 3 days in males wintered for the same period at 0°C (Bosch & Kemp, 2003). Similar results were obtained in *O. cornuta* (Bosch & Kemp, 2004). Respiration rates increase faster and reach emergence values sooner in individuals wintered at 7°C compared with 0°C (Sgolastra, 2007). This physiological response allows *Osmia* populations to advance their emergence in years with mild winters and, therefore, not to miss bloom of their host plants, which also respond to mild winters by advancing their blooming period (Nyéki & Soltész, 1996). However, fat body depletion and mortality increase rapidly at wintering temperatures above 4°C (Bosch & Kemp, 2003). Late-flying *O. lignaria* cannot be wintered for more than 150 days at 7°C. Similarly, *O. cornifrons* can be wintered for more than 180 days at 0°C, but not at 5°C (Maeta et al., 2006).

As with summer diapause, *Osmia* populations appear to be adapted to local winter conditions. In *O. cornuta*, females of February-flying populations emerged within 1.5 days when incubated at 20°C following 120-day wintering at 3°C, whereas April-flying populations exposed to the same treatment required 5 days to emerge (Bosch & Blas, 1994). *Osmia cornuta* can be wintered for 150 days at temperatures as high as 11°C, but the same wintering conditions caused extensive fat body depletion and mortality in *O. lignaria* (Bosch & Kemp, 2003, 2004).

Management of *Osmia* Populations for Crop Pollination

Osmia cornifrons, *O. lignaria*, and *O. cornuta* are very efficient fruit tree pollinators. Densities required for pollination vary from crop to crop and year to year (due in part to alternating blooming intensity), but estimates indicate that 550–750 nesting females (plus 1.5–2 times as many males) per ha usually suffice (Bosch & Kemp, 2002, and references therein). Due to pre nesting dispersal, numbers of females released have to be increased by up to two times depending on the releasing method (see Bosch & Kemp, 2001, 2002). Detailed accounts of management practices for *Osmia* spp. can be found elsewhere (Maeta & Kitamura, 1974; Bosch & Kemp, 2001; Krunic & Stanisavljevic, 2006). Here we discuss only those practices directly related to developmental biology. Due to the important phenological differences among populations discussed in the previous sections, any rearing protocols need to be based on knowledge of the natural phenology in the area of origin of the population under management. Assuming a genetic basis for these phenological differences, it may be possible to select different *Osmia* strains to pollinate crops with different blooming periods, from almonds in February to apples in May.

Monitoring and Managing Emergence Timing

Prior to bloom, *Osmia* nests (or loose cocoons depending on the management system) are taken out of the wintering area and placed in nesting shelters containing nesting materials (e.g., wood blocks with inserted paper straws, bundles of cardboard tubes or reeds). Timing between bee emergence, and therefore nesting, and bloom initiation of the target crop is of the essence when pollinating fruit trees, as most orchards bloom for only about 3 weeks. If bees emerge too late, a high proportion of flowers are not pollinated and, because alternative floral resources are usually scarce after petal fall of the target crop, *Osmia* reproductive success decreases. On the other hand, females emerging too much ahead of bloom initiation are likely to disperse in search of flowers and nest elsewhere.

For this reason, it is very important that populations be wintered for sufficiently long periods, as this will result in short preemergence (time between spring incubation and first emergence) and emergence (time between first and last emergence) periods. Populations with long preemergence periods are more difficult to time with bloom initiation of the target crop. In populations with long emergence periods, correct timing with bloom can only be accomplished with a segment of the population. The need to provide sufficiently long wintering periods to achieve prompt and uniform emergence cannot be overemphasized. In any case, plants blooming ahead of the target crop are convenient pollen-nectar sources for early-emerging *Osmia* individuals (Bosch & Kemp, 2001).

When wintered for long periods at a mild temperature (4–7°C), some males emerge without need of incubation. This is a good indication that a population has received enough wintering to complete diapause and is ready to emerge. Otherwise, male cocoons can be exposed to room temperature (~22°C) and checked for emergence. In a sufficiently wintered population, males emerge within 1 day. As mentioned, wintering periods for prompt emergence vary among species and populations. In April-flying *O. lignaria* populations, some 210 days of wintering at 4°C elicit prompt emergence (mean male preemergence time: 1 day; Bosch & Kemp, 2000, 2003). Instead, February-flying *O. cornuta* emerge promptly following 120 days at 3°C (mean male preemergence time: 1 day; Bosch & Blas, 1994).

Prompt emergence may also be accomplished by wintering bees at warmer temperatures. However, this method should only be used when long wintering periods are not feasible. Increasing winter temperatures can only be done with caution and for short periods of time, because this treatment may result in fat body depletion and loss of vigor. Bosch et al. (2000) wanted to time emergence of an April-flying *O. lignaria* population from northern Utah to February almond bloom in California. The need to obtain emergence by February instead of April drastically reduced the period available for wintering. One part of the population was wintered at 4°C and the other at 7°C for 117 days. Male bees wintered at 4°C took 7 days to emerge compared with 3 days for bees wintered at 7°C, with no significant increase in mortality. Temperature can also be increased gradually throughout wintering to hasten diapause completion and prompt emergence.

Timing of emergence is also greatly influenced by incubation temperature. If wintered for sufficiently long periods, and if maximum daily temperatures are expected to be above 20°C, populations require no artificial incubation. On the other hand, if temperatures are very cold, both flower development and bee emergence are arrested. However, under mild temperatures (10–18°C), fruit tree bloom progresses (Nyéki & Soltész, 1996), and female emergence is very slow. Then it is necessary to provide additional incubation. Nests can be incubated in a temperature chamber at 20–25°C. Emerging adults can be collected into a container and either released in the orchard or cooled to 4–8°C until release time. When using this method, it is important to note that females released as emerged individuals are more likely to disperse and nest elsewhere than females released as nonemerged individuals within their nest (Maeta, 1978; Torchio, 1984; Bosch, 1994c). Nests can also be incubated within the nesting shelter (Bosch & Kemp, 2001; Pitts-Singer, et al., 2008). In any case, release schedules need to account not only for the emergence period, but also for the prenesting period.

Sometimes it may be necessary to delay, rather than advance, emergence. This can be accomplished by cooling populations. *Osmia cornifrons* populations wintered outdoors emerge in synchrony with cherry bloom. To time emergence with apples, which bloom later, emergence is held up by transferring populations to 5°C (Maeta & Kitamura, 1974; Maeta, 1990). However, *Osmia* populations that are ready for emergence cannot be cooled down for periods longer than 1 month without compromising survival and postemergence vigor (Bosch & Kemp, 2001).

Wintering and incubation requirements vary from population to population, and flowering phenology of fruit trees ranges from February (almonds) to May (apples). For these reasons, wintering and incubation schedules need to be adjusted to each particular situation. Decisions on spring incubation regimes and timing of release in the orchard need to be based on knowledge regarding the origin of the population used, close monitoring of flower development of the target crop, and weather forecasts. Flowering periods of species or cultivars blooming ahead of the target crop provide useful phenological references (Maeta, 1990; Bosch et al., 2006).

Monitoring and Managing Development

After petal fall, nesting materials containing progeny are taken to a storage unit with temperatures close to ambient. Development can be easily monitored by periodically dissecting a few male cocoons. As adult eclosion time approaches, monitoring needs to be intensified. To begin with, only male cocoons need to be sampled. Then, as the population approaches 100% eclosion, female cocoons also need to be sampled, because females may sometimes develop more slowly than males. Because eclosion periods last for more than a month, populations should be wintered as soon as all sampled females have reached adulthood. This will prevent early-developing individuals from being exposed to excessively long prewintering periods. In the 1990s, several *Osmia* keepers in the United States reported on populations with low performance (high mortality, low vigor, and poor establishment). These weak populations likely resulted from excessively long prewintering periods. When *O. cornuta* from a population that normally reaches adulthood in September were wintered in late November, emergence was slow and irregular (Bosch, 1994a, 1995). Some individuals of this population died in the process of chewing their way out of the cocoon, and others emerged but were unable to fly. Therefore, monitoring adult eclosion and management of prewintering duration are essential for obtaining healthy, strong *Osmia* populations.

Comparison With *Megachile rotundata*

The alfalfa leafcutting bee, *M. rotundata*, has been established as a crop pollinator in North America since the 1960s (see chapter 7, this volume). Like *Osmia*, *M. rotundata* females nest in preestablished cavities and build linear series of cells provisioned with a mixture of pollen and nectar. However, the life cycle and the wintering physiology of

Osmia and *Megachile* are different, which imposes important differences in management practices. *Megachile* winter in the prepupal stage and fly later in the year than *Osmia*. *Megachile rotundata* is partially bivoltine, that is, within a population some individuals become dormant in the prepupal stage, whereas others continue development into pupae and adults and emerge in late summer to produce a second generation (Krunic, 1972; Richards, 1984; Kemp & Bosch, 2001). Both first- and second-generation individuals winter in the prepupal stage.

Megachile rotundata females lay their eggs in June and July. As in *Osmia*, immatures develop through five instars, and the last instar defecates and spins a cocoon (Trostle & Torchio, 1994). Development from egg to cocoon completion is faster than in *Osmia* (Kemp & Bosch, 2000). As with *Osmia*, *M. rotundata* prepupae enter diapause shortly after the completion of cocoon spinning. Respiration rates drop from 1.3 ml/g·h in the feeding larva to 0.5 at cocoon completion and then to 0.2 within 21 days (Kemp et al., 2004). However, contrary to *Osmia*, respiration rates (measured at 22°C) do not increase with exposure to wintering temperatures. Respiration rates remain at 0.1–0.2 ml/g·h throughout the autumn and winter, instead of increasing steadily as in *Osmia*. In fact, respiration rates do not increase until exposure to incubation temperatures (Kemp et al., 2004). These differences in response to temperature reveal profound physiological differences between the adult winter diapause of *Osmia* and the prepupal winter diapause of *Megachile*.

Megachile rotundata appears to be much more tolerant than *O. lignaria* to long and warm wintering periods. In *M. rotundata* wintered at 5°C, a significant increase in mortality was not observed until wintering periods were extended beyond 270–300 days (Richards, Whitfield, & Schaalje, 1987). In addition, prepupae could be wintered at 10°C for up to 450 days without an apparent increase in mortality. Based on these differences, we would expect *M. rotundata* prepupae to be less sensitive to long prewintering durations than *Osmia* adults. If our hypothesis is correct, the timing of wintering initiation should be less critical in *M. rotundata* than in *Osmia*. Further proof of the physiological differences between prepupal and adult diapause comes from experiments in which *O. cornifrons* prepupae were exposed to cool temperatures for various periods and subsequently exposed to 22°C to complete development (Maeta et al., 2006). Prepupae of this species could be wintered at 6°C for 180 days and at 10°C for 410 days with no increase in mortality.

Overall, the effect of wintering duration on emergence timing follows a similar pattern in *Osmia* and *Megachile*. Preemergence and emergence periods of *M. rotundata* also decrease with increasing wintering duration (Johansen & Eves, 1973; Taséi & Masure, 1978; Richards et al. 1987). The effect of incubation temperatures on emergence is also similar in both genera, but contrary to *O. lignaria*, no detrimental effect of incubation temperatures as high as 32°C has been observed in *M. rotundata* (Richards & Whitfield, 1988). Thus *M. rotundata* populations are routinely incubated at 30°C, and peak female emergence in populations wintered for ~7 months at 5°C occurs within 23–24 days of incubation at 30°C.

Alfalfa blooms for longer periods (~8 weeks) than fruit trees. In addition, the risk of bad weather is much reduced during alfalfa bloom, and not all flowers need to be pollinated to attain maximum seed set (Pedersen, Petersen, Bohart, & Levin, 1956). For these reasons, timing of emergence with bloom initiation is not as critical in *M. rotundata* as it is in *Osmia* management. *Megachile rotundata* populations are released in synchrony with ~10% alfalfa bloom (Richards, 1984). As mentioned, *Osmia* populations need to be released before bloom or when the very first flowers within the orchard are opening.

Concluding Remarks

Basic knowledge about the developmental physiology of bee pollinators is essential for developing appropriate rearing methods. There are important differences in winter diapause development between *Osmia* and *Megachile*, and these differences must be recognized in management protocols. Likewise, phenological differences among populations with different geographical origins need to be considered when establishing rearing protocols for *Osmia*. The development of adult diapause in *Osmia* conforms to the diapause model proposed by Sawyer, Tauber, Tauber, and Ruberson (1993) to explain egg diapause development in the gypsy moth, *Lymantria dispar*. This model assumes that both threshold and optimal temperatures for diapause development change gradually over the course of the winter so that no clear demarcation between diapause and postdiapause exists. Adult *Osmia* respiration rates (measured at 22°C) reach a minimum shortly after adult eclosion and then steadily increase throughout the winter. Early in the diapause period, adults respond to warm temperatures by lowering respiration rates and thus slowing diapause development, whereas cold temperatures at this stage elicit an increase in respiration rates. Instead, at the end of the diapausing period, warm temperatures prompt emergence, and cold temperatures delay it.

Two phenological events require particular attention in *Osmia* management: the timing of adult diapause in the autumn and the timing of emergence in the spring. Monitoring these events is critical to the success of *Osmia* rearing operations. The rest of the management process follows a rather simple routine (see Bosch & Kemp, 2001). In addition, *Osmia* management is logistically simplified due to the small bee densities required for pollination. As already mentioned, a maximum of 1,100–1,500 females need to be released per ha compared with 16,000–40,000 in *M. rotundata* (Hobbs, 1973; Richards, 1984). At these densities, orchard yields can be increased significantly (Bosch & Kemp, 1999; Bosch et al., 2006). In Japan, the use of *O. cornifrons* has increased from 10% of the total apple production area in 1981 to 50% in 1990 to over 70% in 1996 (Maeta, 1990; Sekita et al., 1996; Batra, 1998). Future studies on *Osmia* management should develop phenological models to predict emergence timing as a function of wintering and incubation regimes. These models should be based on female emergence and account for prenesting periods to synchronize maximum female activity with peak bloom. Ideally, separate models should be developed for early- and late-flying strains so as to cover the wide range of blooming periods of different fruit tree species and cultivars.

Acknowledgments

We are grateful to R. R. James, T. L. Pitts-Singer, and P. F. Torchio for their useful comments on a draft of the manuscript.

References

Abel, C. A., Wilson, R. L., & Luhman, R. L. (2003). Pollinating efficacy of *Osmia cornifrons* and *Osmia lignaria* subsp. *lignaria* (Hymenoptera: Megachilidae) on three Bassicaceae grown under field cages. *Journal of Entomological Science, 38,* 545–552.

Asada, S., & Ono, M. (2002). Development of a system for commercial rearing of Japanese native bumblebees, *Bombus hypocrita* and *B. ignitus* (Hymenoptera: Apidae) with special reference to early detection of inferior colonies [in Japanese]. *Japanese Journal of Applied Entomology and Zoology, 46,* 73–80.

Asensio, E. (1984). *Osmia (Osmia) cornuta* Latr. pollinisateur potentiel des arbres fruitiers en Espagne (Hymenoptera, Megachilidae). Fifth International Symposium on Pollination, INRA, Versailles, France, 461–465. INRA (Institut National de la Recherche Agronomique), Paris.

Ayres, P. M., & Scriber, J. M. (1994). Local adaptation to regional climates in *Papilio canadensis* (Lepidoptera: Papilionidae). *Ecological Monographs, 64,* 465–482.

Batra, S. W. T. (1979). *Osmia cornifrons* and *Pithitis smaragdula*, two Asian bees introduced into the United States for crop pollination. *Fourth International Symposium on Pollination*, (Special Miscellaneous Publication 1, 307–312). Maryland Agricultural Experiment Station.

———. (1998). Hornfaced bees for apple pollination. *American Bee Journal, 138,* 364–365.

Bohart, G. E. (1958). Transfer and establishment of the alkali bee. *Proceedings of the Sixteenth Alfalfa Improvement Conference*, Ithaca, New York, 4–6.

———. (1962). How to manage the alfalfa leaf-cutting bee *Megachile rotundata* Fabr. for alfalfa pollination. *Utah Agricultural Experiment Station Circular, 144,* 1–7. Logan, UT.

Bosch, J. (1994a). Improvement of field management of *Osmia cornuta* (Latreille) (Hymenoptera: Megachilidae). *Apidologie, 25,* 71–83.

———. (1994b). The nesting behavior of the mason bee *Osmia cornuta* (Latr) with special reference to its pollinating potential (Hymenoptera: Megachilidae). *Apidologie, 25,* 84–93.

———. (1994c). *Osmia cornuta* Latr. (Hym., Megachilidae) as a potential pollinator in almond orchards: Releasing methods and nest hole-length. *Journal of Applied Entomology, 117,* 151–157.

———. (1995). Comparison of nesting materials for the orchard pollinator *Osmia cornuta* (Hymenoptera, Megachilidae). *Entomologia Generalis, 19,* 285–289.

Bosch, J., & Blas, M. (1994). Effect of overwintering and incubation temperatures on adult emergence in *Osmia cornuta* Latr (Hymenoptera, Megachilidae). *Apidologie, 25,* 265–277.

Bosch, J., & Kemp, W. P. (1999). Exceptional cherry production in an orchard pollinated with blue orchard bees. *Bee World, 80,* 163–173.

Bosch, J., & Kemp, W. P. (2000). Development and emergence of the orchard pollinator *Osmia lignaria* (Hymenoptera: Megachilidae). *Environmental Entomology, 29,* 8–13.

———. (2001). *How to manage the blue orchard bee as an orchard pollinator.* Beltsville, MD: Sustainable Agriculture Network.

———. (2002). Developing and establishing bee species as crop pollinators: The example of *Osmia* spp. (Hymenoptera: Megachilidae) and fruit trees. *Bulletin of Entomological Research, 92,* 3–16.

———. (2003). Effect of wintering duration and temperature on survival and emergence time in the orchard pollinator *Osmia lignaria* (Hymenoptera: Megachilidae). *Environmental Entomology, 32,* 711–716.

———. (2004). Effect of prewintering and wintering temperature regimes on weight loss, survival, and emergence time in the mason bee *Osmia cornuta* (Hymenoptera: Megachilidae). *Apidologie, 35,* 469–479.

———. (2008). Production of undersized offspring in a solitary bee. *Animal Behaviour 75,* 809–816.

Bosch, J., Kemp, W. P., & Peterson, S. S. (2000). Management of *Osmia lignaria* (Hymenoptera: Megachilidae) populations for almond pollination: Methods to advance bee emergence. *Environmental Entomology, 29,* 874–883.

Bosch, J., Kemp, W. P., & Trostle, G. E. (2006). Cherry yields and nesting success in an orchard pollinated with *Osmia lignaria* (Hymenoptera: Megachilidae). *Journal of Economic Entomology, 99,* 408–413.

Bosch, J., Maeta, Y., & Rust, R. (2001). A phylogenetic analysis of nesting behavior in the genus *Osmia* (Hymenoptera: Megachilidae). *Annals of the Entomological Society of America, 94,* 617–627.

Bosch, J., & Vicens, N. (2002). Body size as an estimator of production costs in a solitary bee. *Ecological Entomology, 27,* 129–137.

———. (2005). Sex allocation in the solitary bee *Osmia cornuta*: Do females behave in agreement with Fisher's theory? *Behavioral Ecology and Sociobiology, 59,* 124–132.

———. (2006). Relationship between body size, provisioning rate, longevity and reproductive success in females of the solitary bee *Osmia cornuta*. *Behavioral Ecology and Sociobiology, 60,* 26–33.

Bradshaw, W. E., Holzapfel, C. H., Kleckner, C. A., & Hard, J. J. (1997). Heritability of development time and protandry in the pitcher-plant mosquito, *Wyeomyia smithii*. *Ecology, 78,* 969–976.

Cane, J. H. (2005). Pollination potential of the bee *Osmia aglaia* for cultivated raspberries and blackberries (*Rubus*: Rosaceae). *Hortscience, 40,* 1705–1708.

Crane, E. (1991). *Apis* species of Tropical Asia as pollinators, and some rearing methods for them. *Acta Horticulturae, 288,* 29–48.

Denlinger, D. L. (1991). Relationship between cold hardiness and diapause. In R. E. Lee, Jr., & D. L. Denlinger (Eds.), *Insects at low temperatures* (174–198). New York: Chapman & Hall.

Dogterom, M. H. (1999). Pollination by four species of bees on highbush blueberry. Unpublished doctoral dissertation, Simon Fraser University.

Drummond, F. A., & Stubbs, C. S. (1997). Potential for management of the blueberry bee, *Osmia atriventris* Cresson. *Acta Horticulturae, 446,* 77–83.

Feder, J. L., Roethele, J. B., Wlazlo, B., & Berlocher, S. H. (1997). Selective maintenance of allozyme differences among sympatric host races of the apple maggot fly. *Proceedings of the National Academy of Sciences of the USA, 94*, 11417–11421.

Free, J. B. (1993). *Insect pollination of crops.* London: Academic Press.

Gomi, T. (1997). Geographic variation in critical photoperiod for diapause induction and its temperature dependence in *Hyphantria cunea* Drury (Lepidoptera: Arctiidae). *Oecologia, 111,* 160–165.

Hobbs, G. A. (1973). *Alfalfa leafcutter bees for pollinating alfalfa in western Canada* (Canada Department of Agriculture Publication No. 1495). Ottawa, Ontario.

Hodek, I., & Honěk, A. (1996). *Ecology of Coccinellidae, Series Entomologica 54.* Dordrecht, Netherlands: Kluwer Academic.

Holm, S. N. (1973). *Osmia rufa* L. (Hymenoptera) as a pollinator of plants in greenhouses. *Entomologia Scandinavica, 4,* 217–224.

Johansen, C., Mayer, D., Stanford, A., & Kious, C. (1982). Alkali bees: Their biology and management for alfalfa seed production in the Pacific Northwest (Pacific Northwest Extension Publication No. 155).

Johansen, C. A., & Eves, J. D. (1973). Effects of chilling, humidity and seasonal conditions on emergence of the alfalfa leafcutting bee. *Environmental Entomology, 2,* 23–26.

Kemp, W. P., & Bosch, J. (2000). Development and emergence of the alfalfa pollinator *Megachile rotundata* (Hymenoptera: Megachilidae). *Annals of the Entomological Society of America, 93,* 904–911.

———. (2001). Postcocooning temperatures and diapause in the alfalfa pollinator *Megachile rotundata* (Hymenoptera: Megachilidae). *Annals of the Entomological Society of America, 94,* 244–250.

———. (2005). Effects of temperature on *Osmia lignaria* (Hymenoptera: Megachilidae) prepupa-adult development, survival, and emergence. *Journal of Economic Entomology, 98,* 1917–1923.

Kemp, W. P., Bosch, J., & Dennis, B. (2004). Oxygen consumption during the life cycle of the prepupa-wintering bee *Megachile rotundata* (F.) and the adult-wintering bee *Osmia lignaria* Say (Hymenoptera: Megachilidae). *Annals of the Entomological Society of America, 97,* 161–170.

Krombein, K. V. (1967). *Trap-nesting wasps and bees: Life histories nests and associates.* Washington, DC: Smithsonian Press.

Krunic, M. D. (1972). Voltinism in *Megachile rotundata* (Megachilidae: Hymenoptera) in southern Alberta. *Canadian Entomologist, 104,* 185–188.

Krunic, M. D., & Stanisavljevic, L. (2006). *The biology of the European orchard bee Osmia cornuta.* Belgrade, Serbia: Izdavač.

Ladurner, E., Recla, L., Wolf, M., Zelger, R., & Burgio, G. (2004). *Osmia cornuta* (Hymenoptera: Megachilidae) densities required for apple pollination: A cage study. *Journal of Apicultural Research, 43,* 118–122.

Ladurner, E., Santi, F., Maccagnani, B., & Maini, S. (2002). Pollination of caged hybrid seed red rape with *Osmia cornuta* and *Apis mellifera* (Hymenoptera Megachilidae and Apidae). *Bulletin of Insectology, 55,* 9–11.

Maccagnani, B., Ladurner, E., Santi, F., & Burgio, G. (2003). *Osmia cornuta* (Hymenoptera, Megachilidae) as a pollinator of pear (*Pyrus communis*): Fruit- and seed-set. *Apidologie, 34,* 207–216.

Maeta, Y. (1974). Preliminary report on the utilization of *Osmia cornifrons* for pollination of ladino clover in Japan. *Kontyu, 11*, 4–5. [In Japanese.]

———. (1978). Comparative studies on the biology of the bees of the genus *Osmia* of Japan, with special reference to their managements for pollinations of crops (Hymenoptera: Megachilidae) [in Japanese]. *Tohoku National Agricultural Experiment Station Bulletin No. 57*, 221.

———. (1990). Utilization of wild bees. *Farming Japan, 24*, 13–20.

Maeta, Y., & Kitamura, T. (1974). How to manage the Mame-ko bee (*Osmia cornifrons* Radoszkowski) for pollination of fruit crops. Ask Co Ltd.,

Maeta, Y., Nakanishi, K., Fujii, K., & Kitamura, K. (2006). Exploitation of systems to use a univoltine Japanese mason bee, *Osmia cornifrons* (Radoszkowski), throughout the year for pollination of greenhouse crops (Hymenoptera: Megachilidae). *Chugoku Kontyu, 20*, 1–17.

Maeta, Y., Okamura, S., & Ueda, H. (1990). Mame-ko bachi, *Osmia cornifrons* (Radoszkowski) as a pollinator of blueberries (Hymenoptera: Megachilidae) [in Japanese]. Report of Chugoku Branch, Odokou 32: 33–42.

McWatters, H. G., & Saunders, D. S. (1996). The influence of each parent and geographic origin on larval diapause in the blow fly, *Calliphora vicina. Journal of Insect Physiology, 42*, 721–726.

Michener, C. D. (2000). *The bees of the world*. Baltimore: Johns Hopkins University Press.

Monzón, V. H. (1998). Biología de *Osmia cornuta* Latr. (Hymenoptera, Megachilidae) y su utilización como polinizador de peral (*Pyrus communis*). Unpublished doctoral dissertation, Universitat Autònoma de Barcelona.

Monzón, V. H., Bosch, J., & Retana, J. (2004). Foraging behavior and pollinating effectiveness of *Osmia cornuta* and *Apis mellifera* (Hymenoptera: Megachilidae, Apidae) on 'Comice' pear. *Apidologie, 35*, 575–585.

Mousseau, T. A., & Dingle, H. (1991). Maternal effects in insect life histories. *Annual Review of Entomology, 36*, 511–534.

Nyéki, J., & Soltész, M. (1996). *Floral biology of temperate zone fruit trees and small fruits*. Budapest, Hungary: Akadémiai Kiadó.

O'Toole, C. (2002). *The red mason bee*. Rothley, United Kingdom: Osmia Publications Limited.

Parker, F. D. (1981). A candidate red clover pollinator, *Osmia coerulescens* (L.). *Journal of Apicultural Research, 20*, 62–65.

———. (1989). Nest clustering as a means of managing *Osmia sanrafaelae* (Hymenoptera: Megachilidae). *Journal of Economic Entomology, 82*, 401–403.

Parker, F. D., & Tepedino, V. J. (1982). Maternal influence on diapause in the alfalfa leafcutting bee (Hymenoptera: Megachilidae). *Annals of the Entomological Society of America, 75*, 407–410.

Pedersen, M. W., Petersen, H. L., Bohart, G. E., & Levin, M. D. (1956). A comparison of the effect of complete and partial cross-pollination of alfalfa on pod sets, seeds per pod, and pod and seed weight. *Agronomy Journal, 48*, 177–180.

Pitts-Singer, T. L., Bosch, J., Kemp, W. P., & Trostle, G. E. (2008). Field use of an incubation box for improved emergence timing of *Osmia lignaria* populations used for orchard pollination. *Apidologie, 39*, 235–246.

Ptacek, V. (1989). Nesting strips for *Rhophitoides canus* Ev. (Hymenoptera, Apoidea) in lucerne seed production [in Czech]. *Sbornik Vedeckych Praci, 11*, 261–273.

Richards, K. W. (1984). *Alfalfa leafcutter bee management in Western Canada* (Agriculture Canada Publication No. 1495/E). Ottawa, Ontario: Agriculture Canada.

Richards, K. W., & Whitfield, G. H. (1988). Emergence and survival of leafcutter bees, *Megachile rotundata*, held at constant incubation temperatures (Hymenoptera: Megachilidae). *Journal of Apicultural Research, 27,* 197–204.

Richards, K. W., Whitfield, G. H., & Schaalje, G. B. (1987). Effects of temperature and duration of winter storage on survival and period of emergence for the alfalfa leafcutter bee (Hymenoptera: Megachilidae). *Journal of the Kansas Entomological Society, 60,* 70–76.

Rösner, B. (1994). Chemische Kommunikation bei der Mauerbiene *Osmia rufa* (Megachilidae). Unpublished master's thesis, Universität Wien.

Roth, E. (1990). Erfahrungen mit der Haltung und dem Einsatz der Roten Mauerbiene (*Osmia rufa*) in Kohlbefruchtungsgruppen. *Wissenschaftliche Zeitschrift der Universität Halle, 39,* 11–14.

Sawyer, A. J., Tauber, M. J., Tauber, C. A., & Ruberson, J.R. (1993). Gypsy moth (Lepidoptera: Lymantriidae) egg development: A simulation analysis of laboratory and field data. *Ecological Modelling, 66,* 121–155.

Sekita, N., Watanabe, T., & Yamada, M. (1996). Population ecology of *Osmia cornifrons* (Hymenoptera, Megachilidae) in natural habitats. *Bulletin of the Aomori Apple Experiment Station, 29,* 17–36.

Sgolastra, F. (2007). Ecofisiologia del ciclo biologico di *Osmia lignaria* (Hymenoptera: Megachilidae). Unpublished doctoral dissertation, Università di Bologna.

Southwick, E. E., & Southwick, L., Jr. (1992). Estimating the economic value of honey bees (Hymenoptera: Apidae) as agricultural pollinators in the United States. *Journal of Economic Entomology, 85,* 621–633.

Steffan-Dewenter, I. (2003). Seed set of male-sterile and male-fertile oilseed rape (*Brassica napus*) in relation to pollinator density. *Apidologie, 34,* 227–237.

Stephen, W. P. (1960). Management and renovation of native soils for alkali bee inhabitation (Agricultural Experiment Station Technical Bulletin No. 52, 27–39). Corvallis: Oregon State University.

———. (1962). Propagation of the leaf-cutter bee for alfalfa seed production (Agricultural Experiment Station Bulletin No. 586, 1–16). Corvallis: Oregon State University.

Sugiura, N., & Maeta, Y. (1989). Parental investment and offspring sex ratio in a solitary mason bee, *Osmia cornifrons* (Radoszkowski) (Hymenoptera: Megachilidae). *Japanese Journal of Entomology, 57,* 861–875.

Taséi, J.-N. (1972). Observations preliminaires sur la biologie d'*Osmia (Chalcosmia) coerulescens* L., (Hymenoptera Megachilidae), pollinisatrice de la luzerne (*Medicago sativa* L.). *Apidologie, 3,* 149–165.

Taséi, J.-N., & Masure, M. M. (1978). Sur quelques facteurs influençant le développement de *Megachile pacifica* Panz. (Hymenoptera, Megachilidae). *Apidologie, 9,* 273–290.

Tauber, M. J., Tauber, C. A., & Masaki, S. (1986). *Seasonal adaptations of insects.* New York: Oxford University Press.

Tepedino, V. J., & Torchio, P. F. (1982). Phenotypic variability in the nesting success among *Osmia lignaria propinqua* females in a glasshouse environment (Hymenoptera: Megachilidae). *Ecological Entomology, 7,* 453–462.

Torchio, P. F. (1976). Use of *Osmia lignaria* Say (Hymenoptera: Apoidea, Megachilidae) as a pollinator in an apple and prune orchard. *Journal of the Kansas Entomological Society, 49,* 475–482.

Torchio, P. F. (1984). Field experiments with the pollinator species, *Osmia lignaria propinqua* Cresson (Hymenoptera: Megachilidae) in apple orchards: III. 1977 studies. *Journal of the Kansas Entomological Society, 57,* 517–521.

———. (1985). Field experiments with the pollinator species, *Osmia lignaria propinqua* Cresson in apple orchards: V. (1979–1980), methods of introducing bees, nesting success, seed counts, fruit yields (Hymenoptera: Megachilidae). *Journal of the Kansas Entomological Society, 58,* 448–464.

———. (1989). In-nest biologies and development of immature stages of three *Osmia* species (Hymenoptera: Megachilidae). *Annals of the Entomological Society of America, 82,* 599–615.

———. (1990). *Osmia ribifloris,* a native bee species developed as a commercially managed pollinator of highbush blueberry (Hymenoptera: Megachilidae). *Journal of the Kansas Entomological Society, 63,* 427–436.

———. (2003). The development of *Osmia lignaria* as a managed pollinator of apple and almond crops: A case history. In K. Strickler, & J. H. Cane (Eds.), *For nonnative crops, whence pollinators of the future?* (67–84). Lanham, MD: Entomological Society of America.

Torchio, P. F., & Tepedino, V. J. (1982). Parsivoltinism in three species of *Osmia* bees. *Psyche, 89,* 221–238.

Trostle, G., & Torchio, P. F. (1994). Comparative nesting behavior and immature development of *Megachile rotundata* (Fabricius) and *Megachile apicalis* Spinola (Hymenoptera: Megachilidae). *Journal of Kansas Entomological Society, 67,* 53–73.

van der Steen, J., & de Ruijter, A. (1991). The management of *Osmia rufa* L. for pollination of seed crops in greenhouses. *Proceedings of Experimental and Applied Entomology, 2,* 137–141.

van Heemert, C., de Ruijter, A., van den Eijnde, J. , & van der Steen, J. (1990). Year-round production of bumblebee colonies for crop pollination. *Bee World, 71,* 54–56.

Velthuis, H. H. W., & van Doorn, A. (2006). A century of advances in bumblebee domestication and the economic and environmental aspects of its commercialization for pollination. *Apidologie, 37,* 421–451.

Vicens, N., & Bosch, J. (2000). Pollinating efficacy of *Osmia cornuta* and *Apis mellifera* (Hymenoptera: Megachilidae, Apidae) on "Red Delicious" apple. *Environmental Entomology, 29,* 235–240.

Vicens, N., Bosch, J., & Blas, M. (1993). Análisis de los nidos de algunas *Osmia* (Hymenoptera, Megachilidae) nidificantes en cavidades preestablecidas. *Orsis, 8,* 41–53.

Wajnberg, E., & Hassan, S. A. (1994). *Biological control with egg parasitoids.* Wallingford, United Kingdom: CAB International.

Wei, S.-G., Wang, R., Smirle, M. J., & Xu, H.-L. (2002). Release of *Osmia excavata* and *Osmia jacoti* (Hymenoptera: Megachilidae) for apple pollination. *Canadian Entomologist, 134,* 369–380.

Westrich, P. (1998). *Die Wildbienen Baden-Württembergs.* Stuttgart, Germany: Verlag Eugen Ulmer.

Xu, H.-L., Yang, L.-I., & Kwon, Y. J. (1995). Current status on the utilization of *Osmia* bees as pollinators of fruit trees in China (Hymenoptera: Megachilidae). *Korean Journal of Apiculture, 10,* 111–116.

Yamada, Y., Oyama, N., Sekita, N., Shirasaki, S., & Tsugawa, C. (1971). The ecology of the megachilid bee *Osmia cornifrons* and its utilization for apple pollination [in Japanese]. *Bulletin of the Aomori Apple Experiment Station, 26,* 39–77.

7 Past and Present Management of Alfalfa Bees

Theresa L. Pitts-Singer

Alfalfa (*Medicago sativa* L., Fabaceae) is the third largest crop produced in the United States; nearly 76 million tons of alfalfa and alfalfa mixture were produced in 2005 (National Agricultural Statistics Service, 2006). It is also a large commodity in other countries, including Canada, China, and Argentina. In order to produce alfalfa hay, seed is required. For the alfalfa seed grower, the quantity produced determines the profit gained, especially when seed is grown for a seed company at a set contract price. This chapter examines one key component of alfalfa seed production—the use of bees as specialist pollinators of alfalfa. Two managed bee species that effectively pollinate alfalfa are the alkali bee, *Nomia melanderi* Cockerell (Halictidae), and the alfalfa leafcutting bee, *Megachile rotundata* Fabricius (Megachilidae; Bohart, 1957; Cane, 2002). The history of the commercial implementation of the management of these bee species and how alfalfa seed producers embraced the bees for pollination are portrayed. Use of these solitary bees as commercial pollinators, current and future areas of research, and thoughts on future commercial and academic applications of these bees also are discussed.

A Brief History of *Medicago*

Alfalfa was likely domesticated around Asia Minor about 6,000 years ago to produce food for horses. By 1600, France and Spain were known to grow alfalfa for livestock, and by the eighteenth century, cultivated alfalfa hay provided forage for horses and dairy cattle in northwestern Europe and China. The modern domestic alfalfa variety, *Medicago sativa*, emerged from a diverse assortment of alfalfa cultivars adapted to a wide range of climates in Eurasia. By the mid-seventeenth century, this alfalfa had been brought to the New World by the Spaniards and Portuguese and was being successfully produced. Alfalfa gradually arrived in North America from Mexico and also was transported as "Chilean clover" via ships sailing around Cape Horn to California during the Gold Rush, circa 1850. Before the end of the nineteenth century, under irrigation conditions, alfalfa hay production grew in popularity for raising dairy livestock, expanding into the Great Basin and Rocky Mountains of the United States and then from Nebraska and Kansas to Texas. The eastern United States had acquired alfalfa seed from Europe in the late eighteenth century, but growing conditions (e.g., cold temperatures, inadequate soil minerals, and subsoil drainage) restricted the success of alfalfa production on a broad scale. Eventually, alfalfa became a major crop in the Great Lakes region of the United States. At the turn of the twentieth century, alfalfa was being grown in much of the United States and parts of Canada (Sauer, 1993; Putnam et al., 2001; Russelle, 2001).

By the 1920s alfalfa seed was widely grown in the northwestern United States (Bohart, 1971) and Alberta, Canada (Frank, 2003). Today, alfalfa seed (common and certified) is an important commodity in the northwestern United States, Canada, and China (table 7.1). It is a lesser commodity in Argentina, France, Greece, and Italy (International Seed Federation, 2004). Approximately 80,000 tons of alfalfa seed are marketed worldwide on an annual basis. In Argentina, only certified alfalfa seed is produced. In Europe, certified seed is grown mostly by France and Italy (4,000–5,000 tons in 2003–2004; J.-N. Tasei, personal communication, March 2006). In 2004, over 50% of the seed varieties produced in the United States were proprietary (National Agricultural Statistics Service, 2005).

Alfalfa seed production is enhanced by cross-fertilization (Bohart, 1957), which is largely achieved by bees. The anthers (pollen source) and stigma (pollen receptacle) of an alfalfa flower are exposed only when the flower is "tripped." If a flower is not tripped, then pollination cannot take place. Tripping occurs when the action of a flower-visiting bee releases the pressure on interlocking keel petals, allowing the flower's fused reproductive column to abruptly snap upward from within the keel (Frank, 2003). When the flower is tripped, these reproductive parts strike the bee. Honey bees (*Apis mellifera*) were used in first attempts to pollinate alfalfa with managed bees, but they avoid the "tripping" mechanism of an alfalfa flower. They probe from the side of the flower to extract nectar, rather than probing from the center of the flower, where tripping is induced most effectively. Thus honey bees seldom trip flowers, except in central to

Table 7.1 Alfalfa seed market (tons harvested) (International Seed Federation, 2004)

Production Period	Argentina[a]	Canada	China	France	Greece	Italy	United States	World Total
5-year average	1,458	14,135	12,000	5,670	5,150	4,560	35,000	78,202
2003/2004	1,854	20,388	18,000	5,070	5,200	4,090	31,000	85,799

[a] Certified seed only.

southern California, where hot temperatures are presumed to predispose alfalfa flowers to trip quite readily (G. Maslonka, personal communication, June 2006). The wide success of the alfalfa seed industry in North America, however, was greatly enhanced by the discovery that wild, solitary bees are very effective pollinators in comparison with honey bees.

The prevalent and effective alfalfa-pollinating alkali bees and alfalfa leafcutting bees forage on alfalfa flowers despite the tripping mechanism. Unmanaged populations of these bees were observed visiting cultivated alfalfa in the northwestern United States in the late 1950s (Bohart, 1957). Although there was a time when wild bee species provided sufficient alfalfa pollination, the use of broad-spectrum insecticides and loss of wild lands has diminished the once abundant natural bee populations in agricultural areas (Bohart, 1971; Frank, 2003). Furthermore, the scale of alfalfa seed production and expected yield has increased. Such circumstances have led to the need for commercially managed solitary bees to ensure economically viable alfalfa seed yields.

A Native Alfalfa Pollinator

The alkali bee is native to arid and semiarid regions of the United States, west of the Rocky Mountains (Stephen, 1959; Johansen & Eves, 1973; Hurd, 1979). By the 1950s, seed producers in northwestern states had recognized the pollinating qualities of alkali bees and were promoting bee retention near their alfalfa fields (Bohart, 1950). These growers worked to maintain and expand open landscapes for natural ground-nesting sites and began developing methods for creating artificial sites. In support of these endeavors, researchers characterized the bee's nesting biology and natural history to aid in the commercial development and application of alkali bees for alfalfa seed production. *Nomia melanderi* became the world's only commercially managed ground-dwelling bee species. Bohart (1971) reviewed the early perceptions and use of the alkali bees for alfalfa pollination.

The alkali bee is a solitary bee, meaning that there is no cooperation among adult bees for nest building and brood rearing. However, these bees are strongly gregarious in their nesting habits, excavating nests in soil in close proximity to each other and occurring naturally in large concentrations, averaging approximately 2.5–4.9 million nests/ha (Johansen & Mayer, 1982; Cane, 2003; Stephen, 2003). In artificial alkali bee nest sites, population levels can comprise 13.5–24.7 million nests/ha (Johansen & Mayer, 1982; Stephen, 2003), and pollination of alfalfa by these bees may yield as high as 2,240 kg of clean seed/ha (Johansen & Mayer, 1982). This yield is remarkable when compared with a yield of 168 kg of clean seed/ha where commercial bees are not used (Stephen, 1955, as cited in Bohart, 1957). Alkali bee adults emerge from the soil when temperatures are hot in the late spring or early summer (Johansen & Mayer, 1982). Mating occurs at the nest sites, and females begin nest building soon afterward. Females prefer to use preexisting holes in the ground that are easily excavated. The nest tunnel consists of a vertical shaft (length 30–40 cm, diameter 7–9 mm), terminating in a lateral tunnel with nest cells (length 12–15 mm, diameter 8–9 mm) extending downward from it (figure 7.1A; Batra, 1970; Johansen & Mayer, 1982). Soil initially accumulates around the entrance hole as the female digs the nest tunnel, thereby creating a durable tumulus (heap) at the nest entrance (Johansen & Mayer, 1982). Nest cells are provisioned with nectar and pollen that may be collected from flowers up to a distance of 3.2 km (Bohart, 1950). An egg

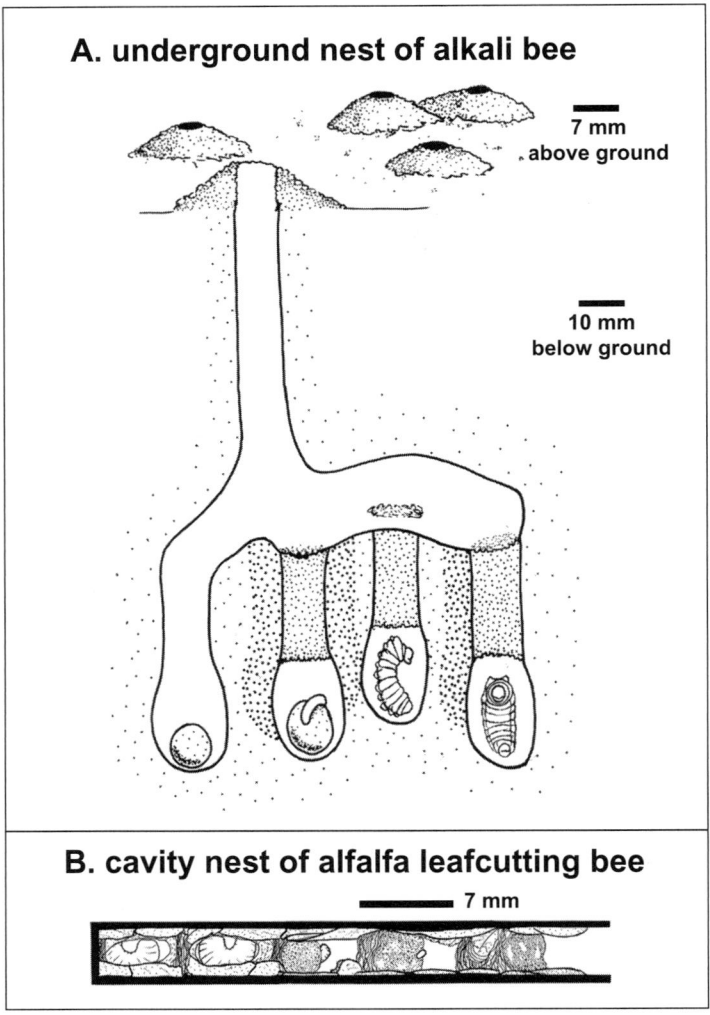

Figure 7.1 Nest morphology of two alfalfa-pollinating bees. (A) Ground nest of *Nomia melanderi* with nest cells containing mass provisions and larvae at various stages of development in separate cells. (B) Cavity nest of *Megachile rotundata* with the oldest larva at the back of the linear series of leaf-lined cells. Relative scale is approximate. Drawing by James P. Pitts (Utah State University, Logan, Utah).

is laid on the provision, and the cell is sealed with a polished soil cap. An adult female can live 4–6 weeks, during which time she may produce 7–12 progeny. The progeny overwinter as prepupae (i.e., postfeeding 5th-instars), and the prepupae complete their transformation through the pupal stage to adults as the soil temperatures warm the following spring and summer. Only one alkali bee generation per year occurs in most of the northwestern United States, but a second generation or more occurs in southern California. Under artificial conditions, increased temperatures may yield up to four generations per year (Stephen, 1965; Johansen & Mayer, 1982).

Meeting the soil requirements for successful nesting is probably the greatest challenge in managing alkali bees. Four basic characteristics of a natural nest site are desired. The soil must be moist, be firm and compact, be bare and sparsely vegetated, and preferably should consist of a silt or fine loam with little or no clay for good capillary movement of water (Johansen & Mayer, 1982; Stephen, 2003). Artificial bee sites often are more productive than managed, natural ones (Johansen & Mayer, 1982), but bringing artificial nest sites up to their reproductive potential takes a few years, as does the recovery of these sites after natural or man-made disasters. Maintaining the nest sites at proper moisture levels seems to be the most critical factor (Johansen & Mayer, 1982).

Managing seminatural alkali bee nest sites may require digging large, water-filled ditches around the sites in order to allow water to seep laterally into the ground. Semiartificial sites may be prepared in deep soil by burying rows of perforated drainpipes through the sites and heavily salting the soil surface. The piping system is used to deliver the recommended approximate 1.9–7.5 million liters per hectare (approximately 200,000–800,000 gallons of water per acre) before the nesting season (Johansen & Mayer, 1982). Early recommendations to create an artificial bee bed were to excavate an area of at least 167 m^2 to a depth of 0.3–0.9 m, to line it with a plastic sheet, and to bury the area under about 30 cm of gravel or coarse sand. Pipes or downspouts were then placed vertically into the gravel to create a system for adding subsurface water to the underlying gravel layer, from which water was expected to move laterally and wick back to the surface. The remainder of the excavation was backfilled with soil and compressed (Johansen & Mayer, 1982).

Today, newer, simpler methods are recommended (M. Wagoner, personal communication, January 2007). Wagoner has established an alkali bee nesting area on his farm, and he describes it as being composed of silty loam and having about a 7% slope. Perpendicular to the slope of the land, Wagoner excavated ditches about 15 cm wide and 51 cm deep, keeping the ditch bottoms level against the slope. The ditches were spaced 1.8 m apart and ran the width of the nest site. He laid water-permeable, weed barrier fabric along the bottom of the ditches before installing 5 cm polyvinyl chloride (PVC) pipes through them. He had previously drilled holes (~1 cm diameter) along one side of these pipes every 30–38 cm to allow water to seep out of the pipes and into the soil. At each end of the pipes, he added 45° elbows for attaching 61 cm upright lengths of pipe so that they stood perpendicular to the ground (standpipes). He covered the ditch pipes with weed barrier fabric and backfilled the ditches with the excavated soil. On one side of the nesting area, Wagoner dug an irrigation ditch from which hoses could draw water; each standpipe was affixed with its own faucet connected to a hose to allow water to be dribbled into each row of pipe. Whether they are semiartificial or newly constructed, salt (NaCl) is added to the surface of nest sites and irrigated into the upper 20 cm of soil, smoothing and sealing the surface.

Alkali bee prepupae can be removed from an established nest site by taking cores of soil containing the ground nests. Thus to establish new nest sites in the early spring, soil cores with prepupae are embedded in trenches cut into the soil. After these transplanted bees emerge, holes are punched into the surface to encourage new nesting, in addition to the reuse of emergence holes as nests (Stephen, 2003). With proper care and maintenance, artificial alkali bee nest sites are known to remain productive for more than 50 years (Cane, 2003).

Protocols developed after 1960 for creating artificial bee nest sites (e.g., Stephen, 1959, 1960; Stephen & Evans, 1960; Johansen & Eves, 1973; Johansen & Mayer, 1982) provided alfalfa seed producers with a system for using a very capable, cost-effective, native pollinator. Many growers in California, Idaho, Nevada, Oregon, Utah, Washington, and Wyoming embraced the use of alkali bees and began to maintain bee nest sites on their property (Mayer & Johansen, 2003), learning how to care for their bees by paying attention to soil moisture and surface salt levels. Awareness of periods of peak bee activity in their regions allowed producers to use various methods to delay the onset of alfalfa bloom, so that the bees were present for the peak flowering period (Johansen & Mayer, 1982). However, the fate of commercial alkali bees was undermined by natural and man-made disasters and by the discovery of a new pollinator, *M. rotundata*, the alfalfa leafcutting bee (Stephen, 2003). The decline in *N. melanderi's* popularity as an alfalfa pollinator is discussed later in this chapter.

A Foreign Alfalfa Pollinator

The alfalfa leafcutting bee, *M. rotundata*, is native to Eurasia and was first collected in the United States in the late 1940s (Hurd, 1954; Stephen, 2003). Although it is not clear how this species came into the United States, its invasive prowess allowed it to successfully establish throughout the country. Once researchers and alfalfa seed producers realized the efficacy of this bee as a pollinator in the late 1950s, its management and propagation flourished (Stephen, 1955; Bohart, 1957; Hobbs, 1965). For example, yields in Canada were only 50 kg clean seed/ha without the use of leafcutting bees, but with the bees, production potential grew to 1,100 kg seed/ha (Richards, 1993).

Like the alkali bee, the alfalfa leafcutting bee is solitary but gregarious and lives for several weeks during the year as an adult. However, this bee species nests in tunnels above the ground (figure 7.1B). Suitable tunnels include insect-chewed cavities in tree trunks or telephone poles, pithy plant stems or hollow twigs, drilled holes in outbuildings or commercial bee boards, and various other natural or man-made cavities of appropriate diameter. Alfalfa leafcutting bees overwinter as prepupae and emerge as adults in the summer when warm temperatures permit completion of development. After mating, nesting female bees use their mandibles to cut circular and oblong leaf pieces from nearby plants, with which they line individual nest cells. Once a cell has been lined, it is provisioned with nectar and pollen, then an egg is laid, and the cell is capped with circular leaf pieces. Another cell is initiated in front of the previous one, until eventually many cells fill the tunnel in a linear series. Fertilized eggs, destined to become females, typically are laid in the innermost cells, and the unfertilized male eggs are laid in the outermost cells. The resulting sex ratio of males:females has been reported as 2:1 (Richards, 1993), although recent work showed that the ratio in commercial populations reared in the United States is closer to 1:1 (Pitts-Singer & James, 2005). Nesting activity continues as the female bee finds another tunnel in which to restart the nest-building process. She may fill several nest tunnels containing 8–12 cells each in her lifetime of 4–6 weeks (Richards, 1984). The development of most progeny temporarily ceases at the prepupal stage in the summer, and they remain in diapause until the next summer. However, some nondiapausing progeny complete development to adulthood to fly during the same growing season as their mothers. The production of adults in the summer (i.e., a "second generation") is an event that does not occur in Old World populations of the alfalfa leafcutting bee (Krunic, 1972; Parker & Tepedino, 1982; Richards, 1984).

Early on, alfalfa seed producers seemed to have been enthused by the ease of attracting alfalfa leafcutting bees for pollination. Alfalfa seed growers in the U.S. Pacific Northwest began a large trap-nesting program in the summer of 1960 (Stephen, 2003). In these early years, growers set out drilled logs, drilled wood boards, and boxes of paper drinking straws and also drilled holes in their outbuildings to encourage bees into the vicinity of their alfalfa fields (Stephen, 1961, 2003; Stephen & Torchio, 1961; Bohart, 1971). Many alfalfa leafcutting bee management innovations also were developed over the subsequent four decades of commercial use (Hobbs, 1967; Stephen, 1981; Richards, 1984; Frank, 2003). Researchers and independent entrepreneurs alike advanced the management and industry of raising alfalfa leafcutting bees through a better understanding of the bees and creating better equipment tailored for their handling and care. Several authors have written descriptions of *M. rotundata* life history and management protocols over the decades, including Stephen (1961, 1962, 1981), Hobbs (1967), Richards (1984), and Frank (2003).

Canadian alfalfa seed producers managed to gain a monopoly on alfalfa leafcutting-bee production (Stephen, 2003), and they are currently the largest suppliers of alfalfa leafcutting bees that are sold as individual cells or in nest-filled bee boards. These producers have developed methods for controlling disease and for generating a surplus of bees, which they vend to U.S. seed producers and, to a lesser extent, to producers in other countries in which alfalfa seed production is undertaken (Donovan et al., 1982; Qingwen et al., 1994; Bitner & Peterson, 2003). The most popular commercial management system is called the "loose cell system," in which the bee cells are removed from the bulky nesting boards. This system allows for removal of debris, parasitoids, and diseased cells and facilitates convenient storage, shipping, and incubation (Baird & Bitner, 1991; Frank, 2003).

The four basic stages of alfalfa leafcutting bee management in the loose cell system are: incubation, release and nesting, removal of loose prepupal cells, and wintering of prepupal cells. Here I describe the system in general, although variations exist between bee managers. The first management stage, *incubation and emergence*, is initiated about 3 weeks before the alfalfa is expected to bloom. Loose bee cells are readied for incubation by placing them in emergence trays with screen tops. Cells in trays are incubated in a chamber that is held at about 30°C, and after about 9 days, dichlorvos resin strips are used to kill any small, parasitic wasps that emerge at this time. The dichlorvos resin strips are removed after about 1 week, and the incubator is ventilated before adult bees emerge. Male bees begin to emerge at about the 20th day of incubation, and females begin to emerge a few days later. After the onset of adult bee emergence, the second stage of management, *field release*, ensues. The trays containing emerged bees are taken to the fields and placed in domiciles.

The domiciles are shelters used for holding bee boards (wood or polystyrene boards with approximately 3,500 holes) that can support thousands of nesting females (figure 7.2). Domiciles are placed within or along the edges of alfalfa fields. The tops of the emergence trays are pulled back or removed so that emerged bees can readily take flight into the fields. The bees will mate, and females will seek out nesting cavities. Once the females find the bee boards, they begin nesting and continue to produce offspring for 4–6 weeks. As they gather nectar and pollen for their offspring, they pollinate the alfalfa flowers. To guarantee some synchrony of bee nesting with flower bloom, some producers will stagger their bee incubation times so that they may release different batches of bees over several weeks. Late in the nesting season, the adult alfalfa leafcutting bees die.

Figure 7.2 Alfalfa leafcutting bee domiciles along the edge of a blooming alfalfa field in Tremonton, Utah. *Photograph by Craig Huntzinger (USDA-ARS, Logan, Utah).*

The third management stage, *harvest of bee progeny*, begins by removing from the field the nest boards containing the progeny and putting the boards into a storage facility where any remaining larvae can develop to the appropriate (i.e., prepupal) stage for overwintering. By the time cold weather sets in, any larvae that have not reached the prepupal stage will die. Eventually, bee cells are removed from the boards. The boards must be dry enough that the cells can be punched or stripped from the nesting material. Cell extraction machines have been invented for this task, having metal rods (sized to fill a tunnel) that are inserted into or alongside the nest tunnels to punch or push the nest contents out of the boards. To avoid damaging cells containing female bees, the boards are oriented so that the outer, presumably male, cells are in direct contact with the rods. Once the bee cells are removed from the nest boards, they first are cleaned in a tumbler that filters out debris and broken cells. Then other machinery is used to break apart cells that remain attached to each other, and some producers further clean their cells using air or gravity separation. The loose cells are stored in containers and refrigerated at about 4°C, which is the fourth stage of management, *wintering*. The bee cells remain in cold storage throughout the winter and are then incubated the following summer so that adult emergence coincides with alfalfa bloom.

Although some European countries produce moderate seed yields compared to production in the United States and Canada (see table 7.1), the alfalfa leafcutting bee is not managed for pollination in Europe as it is in North America (Krunic et al., 1995; J.-N. Tasei, personal communication, March 2006). It is assumed that native bees in Europe's alfalfa seed-growing regions provide adequate pollination for seed production (up to 1,000 kg/ha; Krunic et al., 1995; J.-N. Tasei, personal communication). Moreover, the moist European climate is not as optimal for managing *M. rotundata* as the climate in arid regions of North America (Krunic et al., 1995). The alfalfa seed production acreage in Europe recently has declined dramatically (e.g., seed production in France decreased from 25,000 ha in 1981 to 12,000 ha in 2005), although the average seed yield has increased (e.g., in France increasing from 250 kg/ha in 1981 to 420 kg/ha in 2005; Tasei, 1982; J.-N. Tasei, personal communication). This yield increase may be due to improvement in the pesticides used to combat weeds and alfalfa pests, along with the movement of producers into regions with more suitable seed-growing climates. The natural pollinator fauna appears to be the same as it was 30 years ago; therefore, improved seed production in Europe has not been attributed to better pollination (J.-N. Tasei, personal communication).

Dilemmas Over the Decades

Alkali Bee

The alkali bee was used extensively in the United States into the 1960s, but today very few (< 50) alkali bee nest sites remain, and most are within Walla Walla County in the state of Washington (Cane, 2003; Mayer & Johansen, 2003). Rust (2004, 2005) reports that some active alkali bee nest sites are in Oregon, Idaho, Utah, Colorado, and Nevada, although most of these sites contain few nesting females. It was prophetic that at the time that the excitement over the alkali bee was peaking, Bohart (1950) foresaw its demise. He listed the following three problems: limited species distribution, emergence asynchrony with the first alfalfa blooms, and annihilation by parasites (Bohart, 1950; table 7.2). However, the decline in use of the alkali bee as an alfalfa pollinator also has

Table 7.2 Biological and management-related characteristics affecting the use of two bee pollinators for commercial alfalfa seed production.

Characteristic	Alkali bee	Alfalfa leafcutting bee
1. Nesting behavior	Solitary, but gregarious	Solitary, but gregarious
2. Forage preference	Legumes and more	Legumes and more
3. Pollination proficiency	Excellent	Excellent
4. Natural habitat	South of Canada and west of the Rockies	Native to Eurasia; introduced into Americas and Australia
5. Foraging distance	About 3 miles (4.8 km)	About 1 mile (1.6 mi)
6. Special requirements	Appropriate soil and large area for bee bed; subterranean water and salty soil surface	Nesting materials, field domiciles; cleaning, storage, and incubation equipment; methods for disease and parasite control
7. Management effort	Low, after initial set-up	High throughout
8. Permanence	Permanent, immobile bee nest site	Portable bee cells, nesting material and domiciles
9. Commercial availability	None	Mostly from Canada; also U.S.
10. Production and transfer of surplus progeny	Not easy; use soil cores	Easy, in boards or as loose cells
11. Emergence synchrony; timing with bloom	Depends on locality and natural temperatures; little human control	Depends on incubation timing and temperatures; can stagger releases
12. Estimation of progeny of pollinator population	Difficult for progeny; dig up soil samples for evaluation	Fairly easy with dissection or x-ray of loose cells; rear out some for sex ratio
13. Control of parasitoids and diseases	Few; barriers and pitfall traps for meloid beetles	Yes, chemicals for chalcids and traps for *Sapyga* and beetles; chemicals for chalkbrood disease
14. Used in other crops	No	Yes, canola, melons, carrots and probably others

been attributed to intensified and diversified farming, lack of nest sites, poor weather conditions, direct bee poisoning by crop pesticides, indirect mortality by pesticide contamination of water used to moisten nest sites, and bee disease (Bohart, 1971; Wichelns et al., 1992). Additionally, the current presence of other commercial bees in alfalfa seed stands may pose a negative impact on alkali bees through competition for resources (Mayer & Johansen, 2003).

It is difficult to assess the quality and size of a population of alkali bees from year to year. Being able to predict the population density of female bees assists an alfalfa seed producer in determining whether more bees are needed to fulfill the pollination requirements of the seed crop. A simple but labor-intensive sampling protocol has been devised to estimate reproductive success, as well as mortality from microbial disease, bomber flies, viruses, and other miscellaneous pests (Cane, 2003; Rust, 2005). Although none of the pests and diseases just mentioned are prevalent alkali bee mortality factors (Rust, 2005), some control measures may be taken to combat though not eliminate fungi, bomber flies, blister beetles, and velvet ants (Johansen & Mayer, 1982; Rust, 2005).

Despite the diminished interest in the alkali bee in much of the alfalfa seed-growing regions, use of the bee has not been abandoned completely. Alkali bees are very efficient pollinators that provide consistent pollination across large alfalfa fields (e.g., 160 ha), and not just in the vicinity of the alkali bee nest sites (J. H. Cane, 2007). Today alfalfa seed growers of Walla Walla County continue to maintain their land for alkali bees and depend on these bees for alfalfa seed production (Cane, 2003). Some of these growers supplement their alkali bee pollinator populations with alfalfa leafcutting bees, although this may limit available resources for alkali bee reproduction. As some growers still rely on alkali bees, studies on renovation and establishment of artificial nest sites (Rust, 2006), on bee stocking densities (Cane, 2006), and on associated parasites and diseases (Rust, 2004, 2005, 2006) are warranted.

Furthermore, Rust (2004, 2005) has examined the differences in size, development, emergence, and mortality of alkali bees from different regions of the western United States reared under controlled conditions. This research may lead to a better understanding of adaptations to local climate and to innovations in manipulating or examining factors that would facilitate future commercial use of *N. melanderi*.

Alfalfa Leafcutting Bee

After more than four decades of managing alfalfa leafcutting bees for alfalfa seed production in the United States and Canada, problems remain in sustaining healthy bee populations at the level needed for large-scale pollination. Although early recommendations for the number of bees to release in alfalfa fields were 50,000 bees/ha (Richards, 1984), today's growers inundate their fields with 100,000–150,000 bees/ha (Pitts-Singer & James, 2002–2005, unpublished survey data). In the United States, at the end of the nesting season, the number of healthy progeny produced in alfalfa seed fields is always lower than the number of adult bees released at the beginning of the season. The major culprits for the lack of healthy progeny, in order of magnitude, are the occurrence of "pollen ball" and chalkbrood disease, summer emergence of nondiapausing progeny, and attack by pests and parasitoids (Pitts-Singer & James, 2002–2005, unpublished survey data). The term "pollen ball" is used to describe a cell containing the pollen and nectar provision at a time when the provision should have been consumed by a developing larva. Such cells may account for up to 60% of the cells produced by a population (Bohart, 1971). Examination of these cells reveals that some cells appear to have never had an egg laid in them, that some cells contain collapsed (dead) eggs or larvae, and that some cells are filled with saprophytic fungi (Pitts-Singer, 2004). The cause or causes remain unknown. Proposed theories for the incidence of broodless cells, cells with dead eggs and small larvae, or cell provisions consumed by fungus include the effects of microclimate and the impact of dense bee populations that cause rapid depletion of cell-provisioning materials.

Chalkbrood is a fungal disease of alfalfa leafcutting bee larvae caused by *Ascosphaera aggregata* Skou and is detectable in older larvae (see chapter 8, this volume). *Ascosphaera aggregata* was first detected in the United States in 1975 in a Nevada alfalfa leafcutting bee population. By 1977, 50% of the bee larvae sampled in Nevada were infected, as well as 40% in Idaho and Oregon and 20% in Washington (McManus & Youssef, 1984; Vandenberg & Stephen, 1982). Techniques were developed and are used today in the loose cell system for curtailing the incidence of chalkbrood (Richards, 1984; chapter 8, this volume), yet chalkbrood incidence remains prevalent in the United States.

Nondiapausing alfalfa leafcutting bee progeny that emerge as adults in the same summer in which they were produced are referred to as *second generation.* Although second-generation female bees may add to the pollinator population in the field, they generally die or leave the field because of a lack of bloom, or their offspring die because they do not have time to develop to the overwintering prepupal stage. It is believed that being second generation is a heritable trait (Krunic, 1972; Parker & Tepedino, 1982). However, the fact that most second-generation bees emerge from the cells that are completed early in the season indicates that an environmental component may also be involved (Krunic, 1972). More research is needed to thoroughly understand why and how second-generation emergence occurs.

Parasitoids and other pests can take a heavy toll on alfalfa leaf-cutting bees (Eves et al., 1980; Richards, 1984; Woodward, 1994; Frank, 2003), accounting for more than 20% of the larval mortality at the end of the nesting season (Tasei & Carré, 1982; Pitts-Singer & James, unpublished 2002–2005 survey data). The major pests are small, chalcidoid wasps that may be controlled with light traps and insecticides during the leafcutting bee incubation period, but not during the nesting season. A trap also is available during the nesting season for controlling a common cleptoparasitic wasp, *Sapyga pumila* Cresson (Peterson et al., 1992). The checkered flower beetle, *Trichodes ornatus* Say, lays eggs in bee nests, and its larvae feed on bee provisions, eggs, and larvae. Although a trap with a synthetic lure has been developed for controlling the flower beetle (Davis et al., 1979), I know of no one who uses it. Research on a parasitoid's host acceptance and discrimination would help us to understand the mechanisms involved in hymenopteran parasitism of the alfalfa leafcutting bee, but currently no work is being done on using this information to control these pests (Tepedino, 1988a, 1988b).

Another concern with the alfalfa leafcutting bee is the number of bees needed for pollination. Flooding alfalfa fields with a large number of bees is an approach taken by many U.S. alfalfa seed producers. However, this approach may cause rapid depletion of local floral resources and available nest tunnels, thus decreasing overall bee reproductive success and increasing bee dispersal for alternative resources. Whether this is a good strategy may depend on the grower's desired outcome. U.S. seed producers want to gain quick seed set to avoid early season alfalfa seed pests and maximize their alfalfa seed yield. Flooding a field with bees at peak bloom may be the way to fulfill this desire. In Canada, however, a high bee yield may often be achieved by releasing a more appropriate number of bees for the available resources in areas in which alfalfa seed is a less lucrative commodity.

Finally, the safer use of insecticides during the alfalfa-growing season is a constant consideration. Control of weeds and arthropod pests such as *Lygus* bugs, aphids, alfalfa weevils, and mites demands the use of pesticides for good alfalfa seed yield (Frank, 2003). Johansen and colleagues (1983) and Mayer & Johansen (1999) provide reviews of research done on the various effects of pesticides on bees. Studies in both the laboratory and the field reveal that alfalfa leafcutting bees, alkali bees, bumble bees, and honey bees react differently to pesticides depending on such factors as type of bee exposure, environmental conditions, chemical formulation, dosage and application, and life stage of the bee (Johansen et al., 1983; Frank, 2003; Riedl et al., 2006). Researchers continue to determine effects of new chemicals and formulations on alfalfa leafcutting bee mortality, as well as sublethal effects on foraging activity, nest building, and reproductive success (Barbour et al., 2004).

Future Ventures

The future of alkali bees for commercial pollination of alfalfa is up to those people who are committed to maintaining bee nest sites and providing adequate resources for sustaining bee populations (Mayer & Johansen, 2003). As Wichelns et al. (1992) suggested, changes in the cultural practices associated with crops required to maintain alkali bees might be partly to blame for the bees' decline. Although growers who currently pollinate with alkali bees are satisfied, the alfalfa seed market may not sustain future generations of seed producers who would continue to farm the land and care for the bees. Also, no major companies mass-produce and promote the alkali bee or investigate novel commercial uses for them on crops other than alfalfa. One possibility, however, is the use of this bee for pollination of wildflower seeds, as discussed in chapter 4 in this volume.

It is unfortunate that emphasis on the use of the excellent alfalfa-pollinating alkali bee has waned. Undoubtedly, the success of alfalfa leafcutting bee management for pollination played a major role in the reduction of the use of alkali bees. Agriculture is a business, and producers must seek high yields, even at the expense of high input. The use of alkali bees as an effective pollinator may be more cost-effective once a bee nest site has been established, being less expensive than managing leafcutting bees throughout each year. However, not everyone is able to create and maintain viable nest sites, and not every locality is adaptable for nest sites. In the end, most of today's alfalfa seed producers prefer to use alfalfa leafcutting bees in part because these bees can be managed in areas in which soil and environment are suboptimal for alkali bee nesting. If one compares the commercial characteristics of these two bees, perhaps this preference is justified (see table 7.2). Use of alkali bees as pollinators has advantages over using alfalfa leafcutting bees only in its low input cost after initial establishment and its potential permanence once it becomes productive. All other advantages favor the use of alfalfa leafcutting bees. Until new, more efficient and innovative methods are developed, the demand for alkali bees for alfalfa pollination is unlikely to increase.

In modern North American agriculture, there may be cause for concern over the widespread reliance on the alfalfa leafcutting bee for the alfalfa seed industry. Canada is a primary producer of these bees, and U.S. seed producers are very dependent on this supply. If a catastrophic event were to negatively affect bee supplies (e.g., an uncontrollable disease epidemic or malicious weather conditions during the nesting season), there is no readily available alternative pollinator to replace the alfalfa leafcutting bee. Another possibility is that competition for the supply of leafcutting bees with another crop (e.g., hybrid canola) could drive up the cost of alfalfa leafcutting bees. Higher demand for and cost of bees would impose financial hardship on U.S. alfalfa seed producers but, on the other hand, would benefit Canadian bee producers by boosting the marketability of profitable bees. As another example, consider the problems at hand with the mite *Varroa destructor* in honey bee hives (Doebler, 2000). These mites and the honey bee diseases that they vector are negatively affecting the number of strong hives available for pollination of early spring crops such as almonds in California, leading to high hive rental fees (Sumner & Boriss, 2006) and perhaps less nut production in the near future. A complementary and alternative pollinator is known for almond pollination, the blue orchard bee (see chapter 6, this volume), but a sufficient supply of and protocols for large-scale rearing and management of these bees are not yet available.

Today, alfalfa leafcutting bees are being used as pollinators to hybridize canola seed (Soroka et al., 2001). This species also is effective in pollination of melons (Goerzen & Mueller, 2005) and carrots, especially when they are confined with the crop in enclosures (Tepedino, 1997). Bee producers and researchers will continue to identify new crops that may be pollinated by alfalfa leafcutting bees, and this species may thus become increasingly popular as a pollinator. The effect of increasing demand for alfalfa leafcutting bees to pollinate new crops may ultimately have an impact on the cost and availability of these bees for pollination of alfalfa in the United States. This scenario and its consequences have yet to unfold but are conceivable.

From a nonagricultural standpoint, the alfalfa leafcutting bee may rise in status similar to the honey bee as a model system for studies of insect societies. Many researchers explore the unique characteristics of eusocial insects, but few look at the particular social attributes of solitary insects in light of how such adaptations or behaviors may be evolutionary precursors to eusociality. Studies of solitary aggregating bees that address nest and kin recognition, dominance behavior in nest establishment, nest usurpation, sexual selection, and learning could enhance our understanding of insect social evolutionary history. As scientists reveal the genetic makeup of the honey bee, new tools will become available for comparing genes and gene functions within the Hymenoptera and across insect orders that share similar behaviors. The ease of obtaining and maintaining *M. rotundata* will make them an obvious choice for use in basic evolutionary, behavioral, and genetic studies. From basic research laboratories we can gain a thorough understanding of the alfalfa leafcutting bee that will provide fertile ground for continuing agricultural studies to improve the management of this bee and others.

Acknowledgments

I thank James Cane, Wayne Goerzen, Rosalind James, and James Pitts for their thoughtful comments and careful reviews of this chapter. I also thank Jean-Nöel Tasei and Mark Wagoner for valuable discussions of alfalfa leafcutting bees in Europe and of current practices in making artificial alkali bee nest sites, respectively.

References

Baird, C. R., & Bitner, R. M. (1991). *Loose cell management of alfalfa leafcutting bees in Idaho* (Current Information Series No. 588, 1–4). Moscow: University of Idaho Cooperative Extension.

Barbour, J. D., Gardiner, M. M., & Seymour, L. (2004). University of Idaho alfalfa seed IPM program. In *Proceedings of the Northwest Alfalfa Seed Growers Conference* (49), Reno, Nevada.

Batra, S. W. T. (1970). Behavior of the alkali bee, *Nomia melanderi*, within the nest (Hymenoptera: Halictidae). *Annals of the Entomological Society of America, 63*, 401–406.

Bitner, R. M., & Peterson, S. S. (2003). Introducing the alfalfa leafcutting bee, *Megachile rotundata* (Hymenoptera: Megachilidae), into Australia: A case study. In K. Strickler & J. H. Cane (Eds.), *For nonnativecrops, whence pollinators of the future?* (127–138). Lanham, MD: Entomological Society of America.

Bohart, G. E. (1950). The alkali bee, *Nomia melanderi* Ckll: A native pollinator of alfalfa. In *Proceedings of the 12th Alfalfa Improvement Conference* (32–35), Lethbridge, Alberta.

———. (1957). Pollination of alfalfa and red clover. *Annual Review of Entomology, 2*, 355–380.

———. (1971). Management of habitats for wild bees. *Proceedings of the Tall Timbers Conference on Ecological Animal Control by Habitat Management, 3*, 253–256, Tallahassee, FL: Tall Timbers Research Station.

Cane, J. H. (2002). Pollinating bees (Hymenoptera: Apiformes) of U.S. alfalfa compared for rates of pod and seed set. *Journal of Economic Entomology, 95*, 22–27.

———. (2003). Annual displacement of soil in nest tumuli of alkali bees (*Nomia melanderi*) (Hymenoptera: Apiformes: Halictidae) across an agricultural landscape. *Journal of Kansas Entomological Society, 76*, 172–176.

———. (2006). Feed 'em and reap: Linking bloom, foraging tempos and reproduction by alkali bees. In *Proceedings of the Northwest Alfalfa Seed Growers Conference* (69), Reno, Nevada.

Cane, J. H. (2007). Return of the alkali bee as a commercial pollinator in southeastern Washington. In *Proceedings of the Northwest Alfalfa Seed Growers Conference* (29–30), Las Vegas, Nevada.

Davis, H. G., Eves, J. D., & McDonough, I. M. (1979). Trap and synthetic lure for the checkered flower beetle, a serious predator of alfalfa leafcutting bee. *Environmental Entomology, 8*, 147–149.

Doebler, S. A. (2000). The rise and fall of the honeybee. *Bioscience, 50*, 738–742.

Donovan, B. J., Read, P. E., Wier, S. S., & Griffin, R. P. (1982). Introduction and propagation of the leafcutting bee *Megachile rotundata* (F.) in New Zealand. In *Proceedings of the First International Symposium on Alfalfa Leafcutting Bee Management* (212–222), Saskatoon, Canada: University of Saskatchewan.

Eves, J. D., Mayer, D. F., & Johansen, C. A. (1980). Parasites, predators, and nest destroyers of the alfalfa leafcutting bee, *Megachile rotundata* (Western Regional Extension Publication No. 32). Pullman, WA: U.S. Department of Agriculture and Washington State University Cooperative Extension Service.

Frank, G. (2003). *Alfalfa seed and leafcutter bee: Productions and marketing manual*. Brooks, Alberta, Canada: Irrigated Alfalfa Seed Producers Association.

Goerzen, D. W., & Mueller, S. C. (2005). Alfalfa leafcutting bee alternative crop pollination research in central California. In D. W. Goerzen (Ed.), *Proceedings of the Saskatchewan Alfalfa Seed Producers Association Conference* (19–23), Saskatoon, Saskatchewan, Canada .

Hobbs, G. A. (1965). *Importing and managing the alfalfa leaf-cutter bee* (Publication No. 1209). Ottawa, Ontario, Canada: Canada Department of Agriculture.

———. (1967). Domestication of alfalfa leaf-cutter bees (Canada Department of Agriculture Publication No. 1313). Ottawa, Ontario, Canada: Canada Department of Agriculture.

Hurd, P. D. (1954). Distributional notes on *Eutricharea*, a Palearctic subgenus of *Megachile*, which has become established in the United States (Hymenoptera: Megachilidae). *Entomological News, 65*, 93–95.

———. (1979). Apoidea. In K. V. Krombein, P. D. Hurd, D. R. Smith, & B. D. Burks (Eds.), *Catalog of Hymenoptera in America North of Mexico* (vol. 2, 1741–2209). Washington, DC: Smithsonian Institution Press.

International Seed Federation. (2004). *Forage and turf seed market in selected countries, 2004*. Retrieved March 2006, from the website of the International Seed Federation, http://www.worldseed.org/Statistics/Forage&Turf_2004.htm.

Johansen, C., & Mayer, D. (1982). *Alkali bees: Their biology and management for alfalfa seed production in the Pacific Northwest* (Pacific Northwest Extension Publication No. PWN-155). Pullman: Washington State University.

Johansen, C. A., & Eves, J. (1973). *Management of alkali bees for alfalfa seed production* (Agricultural Cooperative Extension Service Publication No. E.M. 3535). Pullman: Washington State University.

Johansen, C. A., Mayer, D. F., Eves, J. D., & Kious, C. W. (1983). Pesticides and bees. *Environmental Entomology, 12*, 1513–1518.

Krunic, M. D. (1972). Voltinism in *Megachile rotundata* (Megachilidae: Hymenoptera) in southern Alberta. *Canadian Entomologist, 104*, 185–188.

Krunic, M. D., Tasei, J.-N., & Pinzauti, M. (1995). Biology and management of *Megachile rotundata* Fabricius under European conditions. *Apicotura, 10*, 71–97.

Mayer, D. F., & Johansen, C. A. (1999). *How to reduce bee poisoning from pesticides* (Pacific Northwest Extension Publication No. PWN-518). Pullman: Washington State University.

———. (2003). The rise and decline of *Nomia melanderi* (Hymenoptera: Halictidae) as a commercial pollinator for alfalfa. In K. Strickler & J. H. Cane (Eds.), *For nonnative crops, whence pollinators of the future?* (139–149). Lanham, MD: Entomological Society of America.

McManus, W. R., & Youssef, N. N. (1984). Life cycle of the chalkbrood fungus, *Ascosphaera aggregata*, in the alfalfa leafcutting bee, *Megachile rotundata*, and its associated symptomology. *Mycologia, 76,* 830–842.

National Agricultural Statistics Service. (2005). *2004 alfalfa seed production.* Retrieved March 2006 from http://www.nass.usda.gov/mt/pressrls/crops/alfaseed.htm.

———. (2006). *Crop production 2005 summary.* Retrieved March 2006 from the USDA Agricultural Statistics Board website: http://www.nass.usda.gov/wa/agri1oct.pdf.

Parker, F. D., & Tepedino, V. J. (1982). Maternal influence on diapause in the alfalfa leafcutting bee (Hymenoptera: Megachilidae). *Annals of the Entomological Society of America, 75,* 407–410.

Peterson, S. S., Baird, C. R., & Bitner, R. M. (1992). Current status of the alfalfa leafcutting bee, *Megachile rotundata*, as a pollinator of alfalfa seed. *Bee Science, 2,* 135–142.

Pitts-Singer, T. L. (2004). Examination of "pollen balls" in nests of the alfalfa leafcutting bee, *Megachile rotundata*. *Journal of Apicultural Research, 43*(2), 40–46.

Pitts-Singer, T. L., & James, R. R. (2005). Emergence success and sex ratio of commercial alfalfa leafcutting bees, *Megachile rotundata* Say, from the United States and Canada. *Journal of Economic Entomology, 98,* 1785–1790.

Putnam, D., Russelle, M., Orloff, S., Kuhn, J., Fitzhugh, L., Godfrey, L, et al. (2001). *Alfalfa, wildlife and the environment.* Novato, CA: California Alfalfa and Forage.

Qingwen, Z., Richards, K. W., Lou, K., Weiwei, Z., Shaonan, L., Yuzhen, C., et al. (1994). Introduction of alfalfa leafcutter bees (*Megachile rotundata* F.) to pollinate alfalfa in China. *Entomologist, 113,* 63–69.

Richards, K. W. (1984). *Alfalfa leafcutter bee management in Western Canada* (Agriculture Canada Publication No. 1495/E). Ottawa, Ontario: Agriculture Canada.

———. (1993). Non-*Apis* bees as crop pollinators. *Revue Suisse Zoological, 100,* 807–822.

Riedl, H., Johansen, E., Brewer, L., & Barbour, J. (2006). *How to reduce bee poisoning from insecticides* (Pacific Northwest Extension Publication No. PWN-591). Corvallis: Oregon State University.

Russelle, M. P. (2001). Alfalfa: After an 8,000-year journey, the "Queen of Forages" stands poised to enjoy renewed popularity. *American Scientist, 89,* 252–261.

Rust, R. (2004). Offspring production and mortality in the alkali bee, *Nomia melanderi*: Local versus regional patterns. In *Proceedings of the Northwest Alfalfa Seed Growers Conference* (39–40), Reno, Nevada.

———. (2005) Management of alkali bees, *Nomia melanderi* for alfalfa seed production. In *Proceedings of the Northwest Alfalfa Seed Growers Conference* (55–65), Boise, Idaho.

———. (2006). Renovation and establishment of artificial nesting sites for the alkali bee, *Nomia melanderi*. In *Proceedings of the Northwest Alfalfa Seed Growers Conference* (65–67), Reno, Nevada.

Sauer, J. D. (1993). *Historical geography of plants: A select roster.* Boca Raton, FL: CRC Press.

Soroka, J. J., Goerzen, D. W., Falk, K. C., & Bett, K. E. (2001). Alfalfa leafcutting bee (Hymenoptera: Megachilidae) pollination of oilseed rape (*Brassica napus* L.) under isolation tents for hybrid seed production. *Canadian Journal of Plant Science, 81,* 199–204.

Stephen, W. P. (1955). Alfalfa pollination in Manitoba. *Journal of Economic Entomology, 48,* 543–548.

———. (1959). *Maintaining alkali bees for alfalfa seed production* (Agricultural Experiment Station Bulletin No. 568). Corvallis: Oregon State University.

———. (1960). Artificial bee beds for propagation of the alkali bee, *Nomia melanderi. Journal of Economic Entomology, 53,* 1025–1030.

———. (1961). Artificial nesting sites for the propagation of the leaf-cutter bee, *Megachile (Euthicharaea) rotundata,* for alfalfa production. *Journal of Economic Entomology, 54,* 989–993.

———. (1962). *Propagation of the leaf-cutter bee for alfalfa production* (Agricultural Experiment Station Bulletin No. 586). Corvallis: Oregon State University.

———. (1965). Temperature effects on the development and multiple generations in the alkali bee, *Nomia melanderi* Cockerell. *Entomologia Experimentalis et Applicata, 8,* 228–240.

———. (1981). *The design and function of field domiciles and incubators for leafcutting bee management* (Megachile rotundata [*Fabricius*]) (Agricultural Experiment Station Bulletin No. 654). Corvallis: Oregon State University.

———. (2003). Solitary bees in North America agriculture: A perspective. In K. Strickler & J. H. Cane (Eds.), *For nonnative crops, whence pollinators of the future?* (41–66). Lanham, MD: Entomological Society of America.

Stephen, W. P., & Evans, D. D. (1960). *Studies in the alkali bee* (Nomia melanderi *Ckll.*) (Agricultural Experiment Station Bulletin No. 52). Corvallis: Oregon State University.

Stephen, W. P., & Torchio, P. F. (1961). Biological notes on the leaf-cutter bee, *Megachile rotundata (Eutricharaea) rotundata* (Fabricius). *Pan-Pacific Entomology, 37,* 85–93.

Sumner, D. A., & Boriss, H. (2006). Bee-conomics and the leap in pollination fees. *Agricultural and Resource Economics Update, 9,* 9–11.

Tasei, J. N. (1982). Status of *Megachile rotundata* F. in France. In G. H. Rank (Ed.), *Proceedings of the first international symposium on alfalfa leafcutting bee management* (239–246). Saskatoon, Saskatchewan, Canada: University of Saskatchewan Printing Service.

Tasei, J. N., & Carré, S. (1982). Native enemies of *Megachile rotundata* in France. In G. H. Rank (Ed.), *Proceedings of the First International Symposium on Alfalfa Leafcutting Bee Management* (60–64). Saskatoon, Saskatchewan, Canada: University of Saskatchewan Printing Service.

Tepedino, V. J. (1988a). Aspects of host acceptance by *Pteromalus venustus* Walker and *Monodontomerus obsoletus* Fabricius, parasitoids of *Megachile rotundata* (Fabricius), the alfalfa leafcutting bee (Hymenoptera: Chalcididae). *Pan-Pacific Entomologist, 64,* 67–71.

———. (1988b). Host discrimination in *Monodontomerus obsoletus* Fabricius (Hymenoptera: Toryimidae), a parasite of the alfalfa leafcutting bee *Megachile rotundata* (Fabricius) (Hymenoptera: Megachilidae). *Journal of the New York Entomological Society, 96,* 113–118.

———. (1997). A comparison of the alfalfa leafcutting bee (*Megachile rotundata*) and the honey bee (*Apis mellifera*) as pollinators for hybrid carrot seed in field cages: Seventh International Symposium on Pollination. *Acta Horticulturae, 437,* 457–461.

Vandenberg, J. D., & Stephen, W. P. (1982). Etiology and symptomology of chalkbrood in the alfalfa leafcutting bee, *Megachile rotundata*. *Journal of Invertebrate Pathology, 39*, 133–137.

Wichelns, D., Weaver, T. F., & Brooks, P. M. (1992). Estimating the impact of alkali bees on the yield and acreage of alfalfa seed. *Journal of Production Agriculture, 5*, 512–518.

Woodward, D. R. (1994). Predators and parasites of *Megachile rotundata* (F.) (Hymenoptera: Megachilidae), in South Australia. *Journal of Australian Entomological Society, 33*, 13–15.

8   The Problem of Disease When Domesticating Bees

Rosalind R. James

When Colonies Suddenly Collapse

When disease strikes, it can devastate a bee colony or nesting shelter and spread within an entire beekeeping operation. For example, U.S. beekeepers experienced large losses of honey bee colonies during the fall of 2006 and the spring of 2007. Reports vary, but 25% of all commercial colonies may have been lost in one season, with some individual beekeepers losing 75–100% of their bees (as reported by Diana Cox-Foster and Jerry Bromenshenk at the U. S. Department of Agriculture's Colony Collapse Disorder Action Plan Workshop in Beltsville, Maryland, April 24, 2007). These losses followed a similarly large loss of honey bees 2 years previous to this. The deaths were not consistent with any of the known causes of bee die-offs, such as pesticide poisonings, varroa mites, the deadly larval disease called foulbrood, or starvation, which occasionally occurs during the winter. In the absence of any obvious explanation, this new syndrome came to be called colony collapse disorder (often abbreviated CCD), and it may be due to a new virus (Cox-Foster et al., 2007). Why is it that beekeepers experience such massive disease outbreaks when epizootics are seemingly rare in natural populations of bees? I attempt to address this question here, as well as to provide the reader with some broad hypotheses as to what we can do to avoid epizootics (disease outbreaks) in domesticated bees despite the pitfalls associated with trying to medicate bees.

Identifying Infectious Diseases of Bees

Records of diseases in natural populations of bees are scant, but the dearth may be due more to the difficulties we face in finding diseased bees than to the actual rarity of wild-bee diseases. For example, diseased bees are most readily recognized as such when the insect is dead or dying, yet the majority of bee observations and collections focus on actively flying adults or adults making floral visits. Thus our most focused observations may be on the healthiest bees in the population.

One approach to finding what diseases naturally occur in bees is to look for microorganisms associated with either nesting materials or dead bees found in nests. Unfortunately, the microbial techniques required for this are very labor intensive and have major limitations, and so they yield few results despite considerable effort. For example, the methods most frequently used isolate only aerobic microorganisms that are easily cultured in the laboratory on nutrient agar. Aerobic microbes are those that utilize oxygen for growth. Many pathogenic organisms are anaerobic (cannot grow at normal oxygen levels) or require elevated levels of carbon dioxide (CO_2). Pathogenic microbes also sometimes have very unusual nutritional requirements and, as a result, can be difficult to culture; or they may grow much more slowly than many of the common saprophytic microorganisms found in the soil and air. It is these common saprophytic microorganisms that are likely to be found in nests and dead-bee samples. As a result, the saprophytic microorganisms quickly overgrow the pathogens on a Petri plate, and we never know that the pathogens were there. However, these kinds of surveys can tell us about the types of microorganisms commonly associated with bee nests. For example, Batra et al. (1973) conducted a thorough survey of the fungi associated with nests and dead bees for the honey bee (*Apis mellifera mellifera*, Apidae), the alkali bee (*Nomia melanderi*, Megachilidae) and the alfalfa leafcutting bee (*Megachile rotundata*, Megachilidae) in the United States, Costa Rica, and India. Goerzen (1991) and Inglis et al. (1992, 1993) similarly recovered a large number of microorganisms, mainly fungi and bacteria, from alfalfa leafcutting bees and their nests. And Johnson et al. (2005) cultured a variety of fungi from honey bee larvae that had been killed by chalkbrood, a fungal disease caused by *Ascosphaera apis* (Ascomycete; table 8.1). These microbial surveys also may offer opportunities to understand how natural microflora affect the disease dynamics in a nest or colony of bees. For example, the natural microbial flora associated with bees (e.g., those found in the intestinal tract) have been found to affect bee susceptibility to pathogens (Gillespie et al., 2000; Gilliam et al., 1985, 1988). Some of these microbes are, so to speak, the "probiotics" of the bee world.

Knowing these limitations to culturing pathogens from samples of dead bees or nest materials, you can see that an alternative approach is needed to identify pathogens. Another approach has been the use of bioassays. Bioassays have been conducted with bees to test their susceptibility to known insect pathogens, but generally such studies are performed when people are looking for possible nontarget effects of microbial pest control agents. For example, Ball et al. (1994) found that preparations of the fungus *Metarhizium anisopliae* (Deuteromycete) kill honey bees, but only when applied at dosages twentyfold higher than expected to be used for control of locusts, the target pest. Actual field application rates resulted in bee deaths of less than 10%. Lepidoptera-active nuclear polyhedrosis virus (Heinz et al., 1995) and entomopox virus from grasshoppers (Goerzen et al., 1990) were found not to infect bees in bioassays, and thus they are assumed to be nonpathogenic. Goerzen et al. (1990) did find the entomopathogenic

Table 8.1 Examples of infectious diseases in bees

Microbial Group	Pathogen	Known Hosts	Host Stage Affected	Disease
Viruses	Acute bee paralysis virus (ABPV)	Bumble bee (*Bombus terrestris*)	Adult	Acute bee paralysis
		Honey bee (*Apis mellifera*)	Adult	Acute bee paralysis
	Apis iridescent virus	Asian honey bee (*Apis serrana*)	Pupa	Iridescent viral disease
	Deformed wing virus (DWV)	Honey bee (*Apis mellifera*)	Adult	Deformed wing disease
	Kashmir bee virus (KBV)	Honey bee (*Apis mellifera*)	Larva, adult	Kashmir viral disease
Bacteria	*Paenibacillus larvae*	Honey bee (*Apis mellifera*)	Larva	American foulbrood
	Spiroplasma sp.	Bumble bee (*Bombus* spp.),	Adult	Spiroplasmosis May disease
		Honey bee (*Apis mellifera*)	Adult	
	Streptococcus pluton	Honey bee (*Apis mellifera*)	Larva	European foulbrood
Fungi	*Ascosphaera apis*	Honey bee (*Apis mellifera*)	Larva	Chalkbrood
	Ascosphaera aggregata	Alfalfa leafcutting bee (*Megachile rotundata*)	Larva	Chalkbrood
	Ascosphaera torchioi	Blue orchard bee (*Osmia lignaria*)	Larva	Chalkbrood
	Aspergillus flavus	Honey bee (*Apis mellifera*)	Larva	Stone brood
Microsporidia	*Nosema apis*	Honey bee (*Apis mellifera*)	Adult	Dysentery
	Nosema bombi	Bumble bee (*Bombus* spp.)	Adult	Nosema disease

continued

Table 8.1 Contd.

Microbial Group	Pathogen	Known Hosts	Host Stage Affected	Disease
Protozoa	*Apicystis bombi*	Bumble bee (*Bombus terrestris*)	Adult	No name
	Crithidia bombi	Bumble bee (*Bombus terrestris*)	Adult	No name
	Crithidia mellificae	Honey bee (*Apis mellifera*)	Adult	Flagellate disease
	Monoica apis	Honey bee (*Apis mellifera*)	Adult	Gregarine disease

fungus *Beauveria bassiana* (Deuteromycete) to be infectious to larvae of the alfalfa leafcutting bee, causing very high mortality, but I have found that different strains of this fungus can vary greatly in virulence to adult bees. For example, bees treated with 1×10^5 spores had a mortality level that ranged from 30 to 74% for the alfalfa leafcutting bee and from 7 to 42% in the honey bee, depending on the fungal strain tested. Honey bees were always less susceptible to this fungus.

Bioassays are useful for evaluating the virulence or pathogenicity of suspect pathogens and microbial control agents, but they are not as useful for determining what pathogenic diseases naturally occur in bees. Furthermore, it is well known by those working in microbial control of insect pests that insects can be more sensitive to infection in laboratory bioassays than in field situations (e.g., Beegle et al., 1982; Latge et al., 1983; Inglis et al., 1997; Legaspi et al., 1999, 2000; Wraight et al., 2000). The lower levels of infection found in the field are sometimes due to poor survival of the pathogen or other factors that might limit insect exposure to the pathogen. For example, the pathogen may predominate in the field in a location at which the insect is not likely to come into contact with it. All the same, actual exposure and infection rates can be predicted quite accurately (e.g., Kish & Allen, 1978; Ferrandino & Aylor, 1987; Elkinton et al., 1995; Kot et al. 1996; Knudsen & Schotzko, 1999; Long et al., 2000; Klinger et al., 2006).

Low levels of disease in the field can also be a result of environmental conditions that are not conducive to infection. For example, temperature and humidity sometimes inhibit a pathogen's ability to invade the host (e.g., Wallis, 1957; Thompson, 1959; Steinkraus & Slaymaker, 1994; Davidson et al., 2003), or environmental conditions may enhance an insect's ability to resist infection (James & Lighthart, 1992; James et al., 1998). Sometimes it is difficult to distinguish the host's response from the pathogen's. For example, chalkbrood infection levels can be increased by chilling bee larvae that have been exposed to the pathogen (Torchio, 1992; Flores et al., 1996; James, 2005a), but we do not know whether this response is due to a decrease in the immune response in the insect or an increase in the virulence of the pathogen. Regardless, some insects can take advantage of this temperature response by intentionally increasing either individual body or colony temperatures to combat infections, a response that has been called *behavioral fever* (Inglis et al. 1996; Starks et al., 2000).

The most useful information we have on bee diseases comes from experience with managed bees, especially the honey bee, bumble bee (mainly *Bombus terrestris*, Apidae), and alfalfa leafcutting bee. We know that bee disease-causing organisms can come from most all the major groups of microorganisms, including viruses, bacteria, fungi, microsporidia, and protozoa (see table 8.1), similar to the pathogen diversity of other insects. When we keep, or manage, bees, we can monitor the entire life cycle, and if we want to build large populations, we attempt to maximize bee survival rates, often paying attention to the causes of mortality of a particular bee for the first time. Thus it is through this activity that we discover the variety of diseases to which bees are susceptible. In addition, when we domesticate bees, we build up their densities to abnormally high levels, and high population density in itself can increase the prevalence of disease in insects (Anderson & May, 1981; Brown, 1987).

Population density may not be the only determining factor in bee disease outbreaks, however. Social insects already have very concentrated populations yet are not known to be any more susceptible to epizootics than other insects. Why is this? Termites have been shown to be capable of reducing fungal infections by grooming each other to remove fungal spores (Boucias et al., 1996; Rosengaus et al., 2000). Honey bees also demonstrate social behaviors that should decrease disease incidence in the hive, such as the detection and removal of dead and diseased larvae and dead adults (Milne, 1983; Gilliam et al., 1988; Spivak & Reuter, 2001). In addition, worker honey bees clean empty brood cells before the cell is used again for brood rearing, and they defecate outside the hive when weather permits. All of these behaviors should reduce epizootics in the hive, but they are not sufficient to keep out all infectious disease. Stow et al. (2007) found that social bees have greater innate immunity in the form of stronger antimicrobial compounds on the cuticle than do solitary and semisocial bees. In contrast, Evans et al. (2006) found that honey bees have about one-third the number of immune response genes known for houseflies and mosquitoes. They suggest that honey bees either have a few very highly tuned microbial defense systems or rely heavily on hive behaviors to defend against diseases.

Genetic diversity may also play a role in colony susceptibility. New honey bee queens go on one mating flight during their lifetimes, but they may mate with a variable number of drones. Queens who mate with several males (polyandry) produce workers that have more genetic diversity than queens who have mated with only one male. Palmer and Oldroyd (2003) found that this diversity resulted in the colony being more resistant to disease outbreaks. However, some bumble bee species also prefer polyandry, yet the number of males with which the queen mates does not seem to play a critical role in bumble bee immunity (Ruiz-González & Brown, 2006). Why genetic diversity would be more important in one social bee and not the other is not clear. Perhaps it is because honey bees have a much greater colony size than bumble bees or because the immune response in bumble bees may be more generalized and less adapted to specific pathogens. That is, the honey bee immune system may have developed mechanisms for identifying infections caused by specific pathogens, whereas bumble bees respond similarly to the presence of any foreign organism in the body and thus have a less diverse response to pathogen invasion. These ideas are purely speculative at this time.

Most of this discussion has centered on colony-level responses to microbial threats, but also of concern in culturing bees is the prevention of epizootics that spread from one nest to another, be it between honey bee hives, bumble bee colonies, or individual nests of solitary bees. Little is known about disease transmission between colonies, yet this level of disease spread potentially will have the most impact on our ability to culture bees.

The Need for Disease Control Strategies

What happens when we do not concern ourselves with disease control in developing bee management systems? Historically, ignoring the potential for epizootics when trying to newly establish a bee or when moving an established bee to a new continent usually results in good bee production for several years before a disease eventually causes a catastrophe. For example, the alfalfa leafcutting bee is native to Europe and Africa (Friese, 1898; Enkulu, 1988) and was accidentally introduced into the United States (Hurd, 1954). To improve pollination for alfalfa seed production, the alfalfa leafcutting bee was intentionally mass propagated from wild populations in the western United States, starting in the 1960s (Stephen & Torchio, 1961). Unfortunately, no special measures were taken to prevent the concurrent mass propagation of pathogens. When methods were developed for mass nesting and increasing bee populations to very high densities, again, no special measures were taken for preventing epizootics.

During the initial propagations, no diseases were known to occur in this bee. In 1972, a new Ascomycete fungus, *Ascosphaera aggregata*, was found to cause chalkbrood disease in alfalfa leafcutting bees in Spain, and another fungus in the same genus, *A. proliperda*, was found to be causing chalkbrood in a closely related bee, *M. centrus*, in Denmark (Skou, 1975). At approximately the same time, severe outbreaks of an unknown disease were occurring in U.S. managed populations of the alfalfa leafcutting bee. This disease was finally identified as chalkbrood caused by *A. aggregata* (Skou, 1975). Chalkbrood is a fungal disease of honey bees first described in 1916 (Maassen, 1916). In honey bees, this disease is caused by *A. apis*, *A. major* (originally thought to be a subspecies of *A. apis*), or the closely related *Bettsia alvei* (primarily a saprophyte on honey bee pollen stores in the hive). Other *Ascosphaera* spp. have since been identified from chalkbrood-infected bees in managed alfalfa leafcutting bee populations, but the vast majority of infections are, from my personal experience, caused by *A. aggregata*, with *A. proliperda* being the second most common and mostly found in co-infections with *A. aggregata*.

The original sources of *A. aggregata* and *A. proliperda* infections in U.S. populations of the alfalfa leafcutting bee have never been determined, but it is very likely that these pathogens were imported from Europe with the bee. By the time this bee was introduced into Canada, the danger of chalkbrood was well known, and control strategies were implemented. For example, only clean bee stock was introduced; bee cocoons are screened for disease each year, and infected populations destroyed; and sanitation is generally better than in the United States. Today, Canadian beekeepers typically disinfect nesting boards, shelters, and cocoons (Frank, 2003; Hobbs, 2003). As a result, chalkbrood incidence is very low in Canada, less than 5%, and usually less than 1%; whereas disease incidence in the United States currently ranges from 11 to 36% (James & Pitts-Singer, unpublished survey), which is similar to levels found in California in 1977 (Hackett, 1980). In other words, the chalkbrood prevalence in the United States may not have changed much since those first outbreaks!

The first incidence of chalkbrood in the honey bee (caused by *A. apis*) in the United States was documented in California in 1968, and an epizootic occurred in 1970–1971 (Rose & Christensen, 1984). Honey bee chalkbrood was widespread and common in the United States by 1975, but it never caused the pandemic that has occurred with chalkbrood in the alfalfa leafcutting bee, highlighting the fact that the occurrence of an outbreak is not inevitable. Possible reasons for the differences in occurrence of this disease in the two bees are discussed later.

Special Issues Associated With Disease Control in Bees

The most obvious line of defense against disease in managed bee populations is to medicate the insects when they become sick. Unfortunately, bees are not easily medicated, even in the rare cases in which a drug may be available. The first problem to be surmounted is how to treat the bees. Individual bees generally cannot be treated because their numbers are too great, or they may not be amenable to treatment (e.g., larvae may be enclosed in cells in a nest). Barring the ability to treat sick individuals, we can treat whole populations by feeding drugs to the insects or applying medications to the hives. This works well mainly for the social bees: honey bees and bumble bees. For example, antibiotics can be fed in sugar syrup or in powdered sugar placed in the hive to combat foul brood and *Nosema* infections. The beekeeper, however, needs to be careful with the timing of drug applications to ensure that the chemicals do not find their way into honey and other hive products used for human consumption.

Most solitary bees do not readily feed from feeders in the field, and individual adults do not share food with each other (trophallaxis), as do the honey bees. It is also very difficult to treat larvae of solitary bees because they are sequestered inside nests in the soil or in cavities such as twigs, reeds, or holes drilled in boards. It might be possible to apply medications to the flowers being visited by the nesting females as a way to transfer the compounds into the nest provisions. This method was tried by Mayer et al. (1990) to treat the alfalfa leafcutting bee but without consistent success. Getting this method to work depends, in part, on being able to get a sufficient amount of chemical onto the anthers, from which pollen is collected by the bees, or systemically into the pollen or nectar. If a means can be found, an advantage to medicating solitary bees as compared with honey bees is that one does not have to worry about contaminating honey and other human consumables, such as wax destined to be used in cosmetics.

Another ideal approach for disease control is to vaccinate bees. Sadd and Schmid-Hempel (2006) found that the European bumble bee (*B. terrestris*) is able to develop long-term protection against bacterial infection after a sublethal infection, but no true vaccines have ever been developed for bees (or any other invertebrate, as far as I am aware).

A third method of control could be to breed for disease resistance in bees, but unfortunately little is known about immunity in bees and the associated genetics. Some U.S. alfalfa seed growers have tried to breed for chalkbrood resistance by providing no disease control in their populations and never importing bees from Canada or other areas with low levels of chalkbrood. However, I have not been able to confirm whether this approach has worked or not. A similar strategy has been used in honey bees, but mainly for resistance to parasites such as varroa and tracheal mites. This situation should improve in the future. As our understanding of bee genetics and disease immunity improves, so will our ability to successfully breed for a variety of traits.

The Disease Triangle and Epizootics

The options for treating bees are limited, as are the known drugs with which to treat them. Those drugs that are available are borrowed from mammalian health care or plant disease control. However, another approach to the problem does exist, and that is to look at the entire disease cycle, and not just individual sick bees or colonies. I borrow here an old concept from plant pathology, that of the disease triangle (figure 8.1). The basis for the disease triangle is that (1) a pathogenic disease does not consist of just a pathogen, (2) the pathogen must infect a host and cause some deleterious effects on that host before a disease occurs, and (3) the process of the infection and the response of the host are both affected by the environment. Thus a disease consists of three main components: the pathogen, the host, and the environment (Stevens, 1960; figure 8.1). The purpose of the disease triangle is to help us conceptualize the factors that affect epizootics and aid us in preventing disease outbreaks, which should be the main purpose behind a disease control strategy.

Elements of the pathogen that affect epizootics are the microbe's life history traits. For example, we need to ask: What stages initiate infection, when do they occur, and can they be free-living in the environment? How is the pathogen transmitted from one host to another or one colony to another? Is an alternate host required or utilized? Also important are factors such as the virulence of the pathogen and how many propagules are present. The environment, in turn, can affect pathogen survival, virulence, and the ability of the pathogen to produce new infective propagules (figure 8.1).

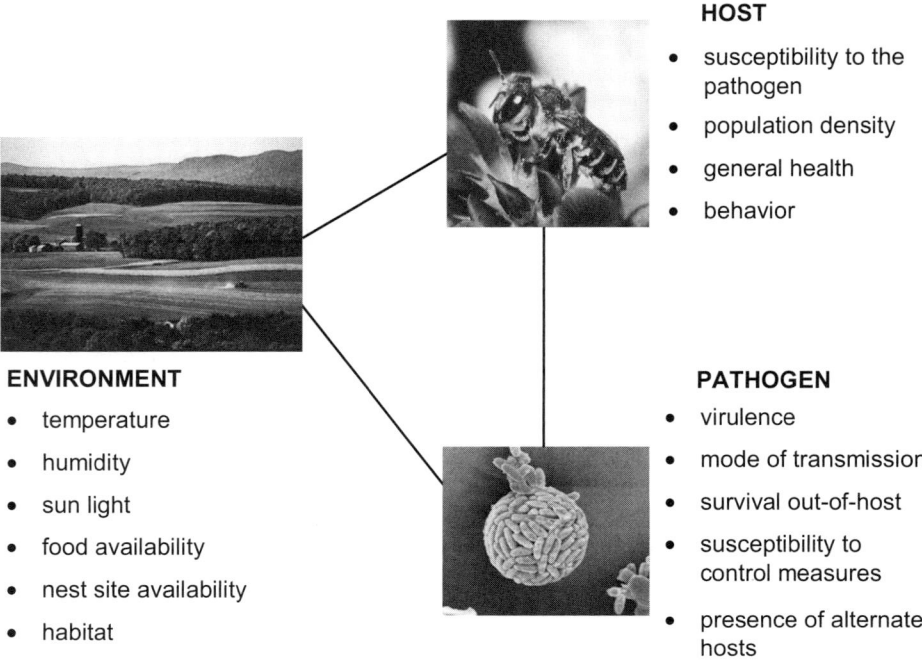

Figure 8.1 The disease triangle for pathogens of bees consists of three elements, the pathogen, the host, and the environment, and the interactions between these elements. Below each element is a list of examples of factors within the element that may have an impact on the manifestation of disease or its prevalence in the host population.

Elements of the host that affect epizootics include: density and abundance, as already discussed; susceptibility to the pathogen; behavior as it affects exposure to the pathogen; life cycle; occurrence of susceptible stages in relation to the occurrence of infective propagules; and genetics (including innate susceptibility). The environment affects growth rate of the host, host susceptibility, and exposure to the pathogen (figure 8.1).

To control a disease, it is important to understand how the components of the disease triangle affect the rate of disease spread in a population over time. Bee managers have little control over many of the host and pathogen factors, but they do have some control over the environment and can manipulate environmental components to favor the host and in this way dispel the disease, if they cannot dispel the pathogen. For example, cool temperature and high moisture favor chalkbrood disease in honey bees, and so this disease can be managed by maintaining hives in areas in which they will not get wet (which may mean using hive stands or screened bottom boards in some climates) and by working to maintain high populations of adult bees so that hive temperature and humidity can be more easily regulated by the colony. When a colony's population drops too low, it may need to be requeened or merged with another colony.

The ability that honey bees have to manipulate environmental conditions may be the primary reason why chalkbrood is not as serious a disease in these bees as it is in alfalfa leafcutting bees. The leafcutting bees are solitary bees and do not regulate nest temperatures. Also, the honey bee pathogen, *A. apis*, does not produce as many infective propagules (spores) in honey bees as *A. aggregata* does in the alfalfa leafcutting bee. It is my observation that leafcutting bees infected with *A. aggregata* are more likely to sporulate than are honey bee larvae infected with *A. apis*. When *A. apis* does sporulate in its honey bee host, often only part of the cadaver produces spores. On the other hand, *A. aggregata* produces spores in the majority of infected alfalfa leafcutting bee larvae, and it is common for the cadaver to be entirely covered in spores. This is, at least in part, an environmental effect (James, 2005a), but other elements of the disease triangle may be important, too.

Sometimes the disease triangle has another dimension, such as a vector for the pathogen. For example, varroa mites have been found to vector several viruses (Chen et al., 2006), and it is probably viruses that cause the severe colony losses found in association with varroa infestations (Martin, 2001). CCD in U.S. honey bees has been correlated with a virus that is also found in varroa-free bees imported to the United States from Australia, where CCD-like symptoms have not been reported to occur. One speculative explanation for the prevalence of CCD in the United States is that varroa depresses the immune system in U.S. honey bees, perhaps making them more susceptible to the virus (Cox-Foster et al., 2007). If this is the case, varroa would add a fourth element to the CCD-disease triangle—not as a vector, but as a second pest that affects the prevalence of the disease.

Adding a vector or a second parasite such as varroa can sometimes simplify disease control and sometimes complicate it. For example, the discovery that mosquitoes serve as the vector for the human pathogen causing yellow fever greatly facilitated the control of this disease. Mosquito abatement programs initiated in Havana in 1901 eliminated yellow fever in the city within the same year. This method was then taken to Panama, where the resulting disease control finally allowed the completion of the Panama Canal, which had been stalled due to the large number of workers that had been contracting malaria (Barrett & Higgs, 2007). However, in the case of varroa, vector control has not been successful. Varroa has proved to be very difficult to control in honey bee hives; thus, although the spread of viruses is greatly enhanced by the vector, this knowledge has not greatly enhanced control of the disease.

Examples of Managing Bee Diseases Through Knowledge of the Disease Cycle

Options for medicating bees are limited, so we need to try to control disease prevalence by breaking the disease cycle in another way. In this section I describe three cases in which this approach has been utilized: American foulbrood in honey bees, *Nosema* infection in bumble bees, and chalkbrood control in the alfalfa leafcutting bee.

American Foulbrood

American foulbrood is a larval disease of honey bees caused by the gram-positive, endospore forming bacterium *Paenibacillus larvae* (see table 8.1). This highly contagious disease spreads easily between colonies on contaminated hive tools and from the reuse of contaminated hive equipment. In the United States, this disease has been controlled with the use of the antibiotic terramycin; however, the pathogen has developed resistance to terramycin, and outbreaks of the disease have since reappeared (Kochansky et al., 2001). Even before antibiotic resistance developed, terramycin did not always entirely eliminate the pathogen from the hive (Wilson et al., 1973), and for this reason, in much of western Europe the disease is controlled instead by destroying hives known to be infected. There, hives are regularly inspected for the disease, and any suspected cases are sent to centralized testing laboratories. Efforts are also made to educate beekeepers in recognizing the disease and avoiding the spread of contaminated materials from one hive to another, including the frequent destruction and replacement of old combs. The success of this program is probably facilitated by the presence of fewer migratory beekeepers in Europe than in the United States.

Nosema *Infection in Bumble Bees*

The second example involves bumble bee rearing and a disease called *nosema*, which is caused by the microsporidian *Nosema bombi*. The Microsporidia were originally classified as Protozoa, but they have recently been reclassified into the Fungi (Keeling, 2002; Adl et al., 2005). In any case, the production of bumble bee colonies can be severely hampered by nosema. This disease causes sublethal infections that start with new queens. Infected bumble bee queens can establish new nests, but the workers that are produced lack vigor and have a reduced life expectancy, and so the growth potential of the colony is compromised. Small colonies also produce fewer new queens in the fall, and infected colonies produce infected queens.

The European bumble bee, *B. terrestris*, has been successfully managed under artificial conditions in Europe, but its importation into the United States has been banned because it is not native to North America. As a result, methods for rearing North American bumble bee species have been pursued. *Bombus occidentals* is a bumble bee native to western North America and has been found to be suitable for domestication, except that it is highly susceptible to nosema (Velthuis & van Doorn, 2006). Another bumble bee that is relatively easy to rear is *B. impatiens*, and it is less susceptible to the disease. Thus *B. impatiens* is the preferred bumble bee for propagation and sale to greenhouse growers in North America. However, this bumble bee does not occur west of the Rocky Mountains, and so its use is restricted in the United States. For example, it can be sold only east of the Rocky Mountains and be used only inside carefully screened greenhouses or other enclosures; "queen-excluders" are required to prevent the accidental release of new queens from the colony into the environment (Velthuis & van Doorn, 2006). The concern is that bumble bees reared in a facility may spread nosema to wild bees. In addition to these measures designed to prevent the spread of the disease from rearing facilities into the environment, control strategies are needed in the rearing facility itself. Like the American foulbrood example earlier, the disease is mainly controlled in rearing facilities by destroying colonies that appear to be infected.

Chalkbrood in the Alfalfa Leafcutting Bee

Both of the preceding examples are extreme cases in which control was achieved by completely destroying colonies in an attempt to control disease spread. In my third example, a different approach was taken. As described earlier, chalkbrood is a serious disease in alfalfa leafcutting bee populations in the western part of North America, where the bee is managed for alfalfa seed production (see chapter 7, this volume). The disease is spread as follows: adult bees become contaminated with spores when they emerge from the natal nests. What happens is that some of the healthy emerging adults get trapped behind dead, diseased siblings in the nest and are forced to chew their way through these cadavers to emerge (Vandenberg et al., 1980; figure 8.2A). Contaminated females then inadvertently deposit the spores with the pollen mass with which they provision their nests (figure 8.2B). The contaminated pollen provisions are then eaten by the developing larvae, who become infected through the gut (Vandenberg & Stephen, 1982; McManus & Youssef, 1984; figure 8.2C), and then die as fully developed larvae, when the fungus finally produces spores (James, 2005a; figure 8.2D).

When I first began working with this bee in 2001, the most common practices used in the United States for controlling the disease were twofold. Most beekeepers removed the bee cells from the nesting boards and then separated the cells from each other mechanically. These cells contain the overwintering cocoons (in a prepupal stage). The idea was to eliminate the source of spore dispersal described previously. Removing the overwintering bees from the nesting boards also allowed the beekeepers to conduct the second step, which was to disinfect the nesting material by baking the boards (for wooden boards) or fumigating them with either methyl bromide or formaldehyde gas (for wooden or polystyrene boards). However, their attempts to disinfect the nesting boards were often not successful. Even when they were successful (or when clean new boards were used), disease levels were unaffected (James, 2005b). Thus, despite these efforts, chalkbrood levels remained the same as before such treatments were implemented.

The explanation for this lack of control was that removing the bee cells from the nesting boards did not eliminate the source of contamination, although that was the intention. The reason is that most chalkbrood-infected larvae die before they spin a cocoon (James, 2005a), and, when the nest cells are separated from the nesting boards and then tumbled to remove debris (another common practice), the cells without cocoons fall apart because they are more fragile. The result is that the spores from the chalkbrood cadavers get spread throughout the batch of bee cells. When the cells are spread out on trays and incubated the next spring, the emerging adults crawl through this contaminated material and pick up large numbers of spores (James & Pitts-Singer, 2005). This source of spores can be targeted by applying fungicides to the cells as they are being prepared for incubation, a method that can reduce chalkbrood incidence in the next

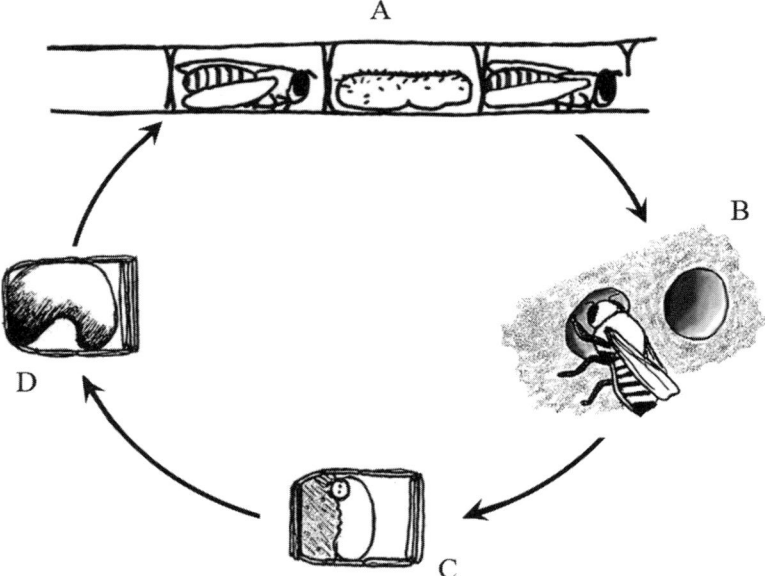

Figure 8.2 Life cycle of chalkbrood in the alfalfa leafcutting bee. (A) Longitudinal cross-section of a nest cavity showing a dead, infected larva (in the center cell) blocking the emergence path of a newly formed adult in the nest. When the bee behind the cadaver emerges, she must chew through the cadaver and, in the process, becomes contaminated with chalkbrood spores. (B) A nesting adult female that is contaminated with spores transfers the spores to the pollen she collects as food provisions for her larvae. (C) A larva contracts the disease after eating spores that contaminate the food provisions in the nest cell. (D) Larvae typically die after they have reached full size, and then the fungus sporulates on the cadaver.

generation (James, 2006). The Canadians also control this source of spores by fumigating the loose cells with paraformaldehyde (Goerzen, 1993; Goettel & Duke, 1996), and they have a very low incidence of chalkbrood, as described earlier.

An important point here is that initial efforts to control the disease focused on disinfection of the nesting materials, a practice that missed targeting a major mode of transmission for this disease. A similar error may be occurring in our attempts to control foulbrood in the honey bee. Dobbelaere et al. (2001) found that hive disinfection methods that eliminate > 99.9% of the *P. larvae* spores do not control foulbrood disease. Perhaps other forms of disease transmission play a key role in persistence of this disease, such as healthy honey bee colonies robbing diseased ones (Fries & Camazine, 2001).

Summary

As with all other biota on Earth, bees are susceptible to diseases, and when we domesticate them, we increase their population densities to suit our needs. This increase in population density makes them more susceptible to disease, and it also brings the natural diseases to our attention when we see them for the first time. Thus, when we domesticate bees, we need to pay attention to disease control. Most examples of bee disease control are known from honey bees because of our long history of managing this bee. Most modern attempts at honey bee disease control have centered on either the use of drug treatments or the destruction of entire colonies and all the associated hive equipment. However, it is clear that disease control strategies are more likely to be successful when the disease cycle within the managed system is understood and taken into account. As with the discovery and implementation of integrated pest management systems for pest control in crops, we are likely to obtain the most utilitarian disease control if we avoid thinking that the only means of controlling disease is being sanitary and destroying infected colonies. We must make an attempt to understand how the disease is spread, identify that time and place in our management system for which the pathogen is most vulnerable, then target our treatments toward that stage, be the treatments chemical, biological, genetic, or otherwise.

Acknowledgments

I am grateful to Jørgen Eilenberg, Jeffrey Pettis, and Theresa Pitts-Singer for comments on earlier drafts of this chapter.

References

Adl, S. M., Simpson, G. B., Farmer, M. A., Andersen, R. A., Anderson, O. R., Barta, J. R., et al. (2005). The new higher level classification of eukaryotes with emphasis on the taxonomy of protists. *Journal of Eukaryototic Microbiology, 52,* 399–451.

Anderson, R. M., & May, R. M. (1981). The population dynamics of microparasites and their invertebrate hosts. *Journal of Animal Ecology, 291,* 451–524.

Ball, B. V., Pye, B. J., Carreck, N. L., Moore, D., & Bateman, R. P. (1994). Laboratory testing of a mycopesticide on non-target organisms: The effects of an oil formulation of *Metarhizium flavorviride* applied to *Apis mellifera*. *Biological Control Science and Technology, 4,* 289–296.

Barrett, A. D. T., & Higgs, S. (2007). Yellow fever: A disease that has yet to be conquered. *Annual Review of Entomology, 52,* 209–229.

Batra, L. R., Batra, S. W. T., & Bohart, G. E. (1973). The mycoflora of domesticated and wild bees (Apoidea). *Mycopathologia et Mycologia Applicata, 49,* 13–44.

Beegle, C. C., Dulmage, H. T., & Wolfenbarger, D. A. (1982). Relationships between laboratory bioassay-derived potencies and field efficacies of *Bacillus thuringiensis* isolates with different spectral activities. *Journal of Invertebrate Pathology, 39,* 138–148.

Boucias, D. G., Stokes, C., Storey, G., & Pendland, J. C. (1996). The effects of imidacloprid on the termite *Reticulitermes flavipes* and its interaction with the mycopathogen *Beauveria bassiana*. *Pflanzenschutz-Nachrichten Bayer* [English ed.], *49,* 103–144.

Brown, G. (1987). Modelling. In J. R. Fuxa (Ed.), *Epizootiology of insect diseases* (43–68). New York: Wiley.

Chen, Y., Evans, J., & Feldlaufer, M. (2006). Horizontal and vertical transmission of viruses in the honey bee, *Apis mellifera. Journal of Invertebrate Pathology, 92,* 152–159.

Cox-Foster, D. L., Conlan, S., Holmes, E. C., Palacios, G., Evans, J. D., et al. (2007). A metagenomic survey of microbes in honey bee colony collapse disorder. *Science, 318,* 283–287.

Davidson, G., Phelps, K., Sunderland, K. D., Pell, J. K., Ball, B. V., Shaw, K., et al. (2003). Study of temperature-growth interactions of entomopathogenic fungi with potential for control of *Varroa destructor* (Acari: Mesostigmata) using a nonlinear model of poikilotherm development. *Journal of Applied Microbiology, 94,* 816–825.

Dobbelaere, W., de Graaf, D. C., Reybroeck, W., Desmedt, E., Peeters, J. E., & Jacobs, F. J. (2001). Disinfection of wooden structures contaminated with *Paenibacillus larvae* subsp. *larvae* spores. *Journal of Applied Microbiology, 91,* 212–216.

Elkinton, J. S., Dwyer, G., & Sharov, A. (1995). Modeling the epizootiology of gypsy moth nuclear polyhedrosis virus. *Computers and Electronics in Agriculture, 13,* 91–102.

Enkulu, L. (1988). Les *Megachiles* (Hymenoptera, Apiodea) d'Europe et d'Afrique. Doctoral dissertation, Faculté des Sciences Agronomiques de L'Etat, Gembloux, Belgium.

Evans, J. D., Aronstein, K., Chen, Y., Hetru, C., Imlert, J.-L., Jiang, H., et al. (2006). Immune pathways and defense mechanisms in honey bees *Apis mellifera. Insect Molecular Biology, 15,* 645–656.

Ferrandino, F. J., & Aylor, D. E. (1987). Relative abundance and deposition gradients of clusters of urediniospores of *Uromyces phaseoli. Phytopathology, 77,* 107–111.

Flores, J. M., Ruiz, J. A., Ruz, J. M., Puerta, F., Bustos, M., Padilla, F., et al. (1996). Effect of temperature and humidity of sealed brood on chalkbrood development under controlled conditions. *Apidologie, 27,* 185–192.

Frank, G. (2003). *Alfalfa seed and leafcutter bee productions and marketing manual.* Brooks, Alberta, Canada: Irrigated Alfalfa Seed Producers Association.

Fries, I., & Camazine, S. (2001). Implications of horizontal and vertical pathogen transmission for honey bee epidemiology. *Apidologie, 32,* 199–214.

Friese, H. (1898). *Die Bienen Europa's (Apidae eropaeae). Theil V. Solitare Apiden: Genus Lithurgus Genus Megachile (Chalkbicodoma).* Innsbruck, Austria: Durck Lampe.

Gillespie, J. P., Bailey, A. M., Cobb, B., & Vilcinskas, A. (2000). Fungi as elicitors of insect immune responses. *Archives of Insect Biochemistry and Physiology, 44,* 49–68.

Gilliam, M., Buchman, S. L., & Lorenz, B. J. (1985). Microbiology of the larval provisions of the stingless bee, *Trigona hypogea,* and obligate necrophage. *Biotropica, 17,* 28–31.

Gilliam, M., Taber, S. I., Lorenz, B. J., & Prest, D. B. (1988). Factors affecting development of chalkbrood disease in colonies of honey bees, *Apis mellifera,* fed pollen contaminated with *Ascosphaera apis. Journal of Invertebrate Pathology, 52,* 314–325.

Goerzen, D. W. (1991). Microflora associated with the alfalfa leafcutting bee, *Megachile rotundata* (Fab) (Hymenoptera: Megachilidae) in Saskatchewan, Canada. *Apidologie, 22,* 553–561.

———. (1993). *Paraformaldehyde fumigation for surface decontamination of alfalfa leafcutting bee cells* (SASPA Extension Publication No. 93–01-2). Saskatchewan, Canada: Saskatchewan Alfalfa Seed Producers Association.

Goerzen, D. W., Erlandson, M. A., & Moore, K. C. (1990). Effect of two insect viruses and two entomopathogenic fungi on larval and pupal development in the alfalfa leafcutting bee, *Megachile rotundata* (Fab.) (Hymenoptera: Megachilidae). *Canadian Entomologist, 122,* 1039–1040.

Goettel, M. S., & Duke, G. M. (1996). Decontamination of *Ascosphaera aggregata* spores from alfalfa leafcutting bee cadavers and bee cells by fumigation with paraformaldehyde. *Bee Science, 4,* 26–29.

Hackett, K. J. (1980). A study of chalkbrood disease and viral infection of the alfalfa leafcutting bee. Unpublished doctoral dissertation, University of California, Davis.

Heinz, K. M., McCutchen, B. F., Herrmann, R., Parrella, M. P., & Hammock, B. D. (1995). Direct effects of recombinant nuclear polyhedrosis viruses on selected nontarget organisms. *Journal of Economic Entomology, 88,* 259–264.

Hobbs, W. (2003). Leafcutter bee chalkbrood research and management. *Forage Seed News, 10,* 27–28.

Hurd, P. D., Jr. (1954). Distributional notes on *Eutricharea*, a Palearctic subgenus of *Megachile*, which has become established in the United States (Hymenoptera: Megachilidae). *Entomological News, 65,* 93–95.

Inglis, G. D., Johnson, D. L., & Goettel, M. S. (1996). Effects of temperature and thermoregulation on mycosis by *Beauveria bassiana* in grasshoppers. *Biological Control, 7,* 131–139.

Inglis, G. D., Johnson, D. L., & Goettel, M. S. (1997). Field and laboratory evaluation of two conidial batches of *Beauveria bassiana* (Balsamo) Vuillemin against grasshoppers. *Canadian Entomologist, 129,* 171–186.

Inglis, D. G., Sigler, L., & Goettel, M. S. (1992). *Trichosporonoides megachiliensis*, a new hyphomycete associated with alfalfa leafcutter bees, with notes on *Trichosporonoides* and *Moniliella*. *Mycologia, 84,* 555–570.

———. (1993). Aerobic microorganisms associated with alfalfa leafcutter bees (*Megachile rotundata*). *Microbial Ecology, 26,* 125–143.

James, R. R. (2005a). Temperature and chalkbrood development in the alfalfa leafcutting bee. *Apidologie, 36,* 15–23.

———. (2005b). Impact of disinfecting nesting boards on chalkbrood control in the alfalfa leafcutting bee. *Journal of Economic Entomology, 98,* 1094–1100.

———. (2006). Testing the use of fungicide applications to loose cells for chalkbrood control. In *Proceedings of the Northwest Alfalfa Seed Growers Conference* (43–44), Reno, Nevada.

James, R. R., Croft, B. A., Shaffer, B. T., & Lighthart, B. (1998). Impact of temperature and humidity on host-pathogen interactions between *Beauveria bassiana* and a coccinellid. *Environmental Entomology, 27,* 1506–1513.

James, R. R., & Lighthart, B. (1992). The effect of temperature, diet, and larval instar on the susceptibility of an aphid predator, *Hippodamia convergens* (Coleoptera: Coccinellidae), to the weak bacterial pathogen *Pseudomonas fluorescens*. *Journal of Invertebrate Pathology, 60,* 215–218.

James, R. R., & Pitts-Singer, T. L. (2005). *Ascosphaera aggregata* contamination on alfalfa leafcutting bees in a loose cell incubation system. *Journal of Invertebrate Pathology, 89,* 176–178.

Johnson, R. N., Zaman, M. T., Decelle, M. M., Siegel, A. J., Tarpy, D. R., Siegel, E. C., et al. (2005). Multiple micro-organisms in chalkbrood mummies: Evidence and implications. *Journal of Apicultural Research, 44,* 29–32.

Keeling, P. J. (2002). Congruent evidence from α-tubulin and β-tubulin gene phylogenies for a zygomycete origin of microsporidia. *Fungal Genetics and Biology, 38,* 298–309.

Kish, L. P., & Allen, G. E. (1978). The biology and ecology of *Nomuraea rileyi* and a program for predicting its incidence on *Anticarsia gemmatalis* in soybean. Unpublished doctoral dissertation, University of Florida.

Klinger, E., Groden, E., & Drummond, F. (2006). *Beauveria bassiana* horizontal infection between cadavers and adults of the Colorado potato beetle, *Leptinotarsa decemlineata* (Say). *Environmental Entomology, 35,* 992–1000.

Knudsen, G. R., & Schotzko, D. J. (1999). Spatial simulation of epizootics caused by *Beauveria bassiana* in Russian wheat aphid populations. *Biological Control, 16,* 318–326.

Kochansky, J., Knox, D. A., Feldlaufer, M., & Pettis, J. S. (2001). Screening alternative antibiotics against oxytetracycline-susceptible and -resistant *Paenibacillus larvae. Apidologie, 32,* 215–222.

Kot, M., Lewis, M. A., & van den Driessche, P. (1996). Dispersal data and the spread of invading organisms. *Ecology, 77,* 2027–2042.

Latge, J. P., Silvie, P., Papierok, B., Remaudiere, G., Dedryver, C. A., & Rabasse, J. M. (1983). Advantages and disadvantages of *Conidiobolus obscurus* and of *Erynia neoaphidis* in the biological control of aphids. In R. Cavalloro (Ed.), *Aphid Antagonists: Proceedings of a Meeting of the EC Experts'Group* (20–32). Rotterdam, Netherlands: Balkema.

Legaspi, J. C., Legaspi, B. C., & Saldaa, R. R. (1999). Laboratory and field evaluations of biorational insecticides against the Mexican rice borer (Lepidoptera: Pyralidae) and a parasitoid (Hymenoptera: Braconidae). *Journal of Economic Entomology, 92,* 804–810.

Legaspi, J. C., Poprawski, T. J., & Legaspi, B. C. (2000). Laboratory and field evaluation of *Beauveria bassiana* against sugarcane stalkborers (Lepidoptera: Pyralidae) in the lower Rio Grande valley of Texas. *Journal of Economic Entomology, 93,* 54–59.

Long, D., Drummond, F., Groden, E., & Donahue, D. (2000). Modelling *Beauveria bassiana* horizontal transmission. *Agricultural and Forest Entomology, 2,* 19–32.

Maassen, A. (1916). Über bienenkrankheiten. *Mitteilungen Kaiserlichen Biologischen Anstalt für Land-und Forstwirtschaft, 16,* 51–58.

Martin, S. J. (2001). The role of *Varroa* and viral pathogens in the collapse of honeybee colonies: A modeling approach. *Journal of Applied Ecology, 38,* 1093.

Mayer, D. F., Lunden, J. D., & Miliczky, E. R. (1990). Effects of fungicide on chalkbrood disease of alfalfa leafcutting bee. *Applied Agricultural Research, 5,* 223–226.

Milne, C. P. (1983). Honey bee (Hymenoptera: Apidae) hygienic behavior and resistance to chalkbrood. *Annals of the Entomological Society of America, 76,* 384–387.

McManus, L., & Youssef, N. N. (1984). Life cycle of the chalk brood fungus, *Ascosphaera aggregata,* in the alfalfa leafcutting bee, *Megachile rotundata,* and its associated symptomology. *Mycologia, 76,* 830–842.

Palmer, K. A., & Oldroyd, B. P. (2003). Evidence for intra-colonial variance in resistance to American foulbrood of honey bees (*Apis mellifera*): Further support for the parasite/pathogen hypothesis for the evolution of polyandry. *Naturwissenschaften, 90,* 265–268.

Rose, J. B., & Christensen, M. (1984). *Ascosphaera* species inciting chalkbrood in North America and a taxonomic key. *Mycotaxon, 19,* 41–55.

Rosengaus, R. B., Traniello, J. F. A., Lefebvre, M. L., & Carlock, D. M. (2000). The social transmission of disease between adult male and female reproductives of the dampwood termite *Zootermopsis angusticollis*. *Ethology Ecology and Evolution, 12*, 419–433.

Ruiz-González, M. X., & Brown, M. J. F. (2006). Males vs. workers: Testing the assumptions of the haploid susceptibility hypothesis in bumblebees. *Behavioral Ecology and Sociobiology, 60*, 501–509.

Sadd, B. M., & Schmid-Hempel, P. (2006). Insect immunity shows specificity in protection upon secondary pathogen exposure. *Current Biology, 16*, 1206–1210.

Skou, J. P. (1975). Two new species of *Ascosphaera* and notes on the conidial state of *Bettsia alvei*. *Friesia, 11*, 62–74.

Spivak, M., & Reuter, G. S. (2001). Resistance to American foulbrood disease by honey bee colonies *Apis mellifera* bred for hygienic behavior. *Apidologie, 32*, 555–565.

Starks, P. T., Blackie, C. A., & Seeley, T. D. (2000). Fever in honeybee colonies. *Naturwissenschaften, 87*, 229–231.

Steinkraus, D. C., & Slaymaker, P. H. (1994). Effect of temperature and humidity on formation, germination, and infectivity of conidia of *Neozygites fresenii* (Zycomycetes: Neozygitaceae) from *Aphis gossypii* (Homoptera: Aphididae). *Journal of Invertebrate Pathology, 64*, 130–137.

Stephen, W. P., & Torchio, P. F. (1961). Biological notes on the leafcutter bee, *Megachile (Eutricharaea) rotundata* (Fabricius). *Pan-Pacific Entomologist, 32*, 84–93.

Stevens, R. B. (1960). Cultural practices in disease control. In J. G. Horsfall & A. E. Dimond (Eds.), *Plant pathology: An advanced treatise* (Vol. 3, pp. 357–429). New York: Academic Press.

Stow, A., Briscoe, D., Gillings, M., Holley, M., Smith, S., Leys, R., et al. (2007). Antimicrobial defenses increase with sociality in bees. *Biology Letters, 3*, 422–424.

Thompson, G. C. (1959). Thermal inhibition of certain polyhedrosis virus diseases. *Journal of Invertebrate Pathology, 1*, 189–192.

Torchio, P. F. (1992). Effects of spore dosage and temperature on pathogenic expressions of chalkbrood syndrome caused by *Ascosphaera torchioi* within larvae of *Osmia lignaria propinqua* (Hymenoptera: Megachilidae). *Environmental Entomology, 21*, 1086–1091.

Vandenberg, J. D., Fichter, B. L., & Stephen, W. P. (1980). Spore load of *Ascosphaera* species on emerging adults of the alfalfa leafcutting bee, *Megachile rotundata*. *Applied and Environmental Microbiology, 39*, 650–655.

Vandenberg, J. D., & Stephen, W. P. (1982). Etiology and symptomology of chalkbrood in the alfalfa leafcutting bee, *Megachile rotundata*. *Journal of Invertebrate Pathology, 39*, 133–137.

Velthuis, H. H. W., & van Doorn, A. (2006). A century of advances in bumblebee domestication and the economic and environmental aspects of its commercialization for pollination. *Apidologie, 37*, 421–451.

Wallis, R. C. (1957). Incidence of polyhedrosis of gypsy-moth larvae and the influence of relative humidity. *Journal of Economic Entomology, 50*, 580–583.

Wilson, W. T., Elliot, J. R., & Hitchcock, J. D. (1973). Treatment of American foulbrood with antibiotic extender patties and antibiotic paper packs. *American Bee Journal, 112*, 341–344.

Wraight, S. P., Carruthers, R. I., Jaronski, S. T., Bradley, C. A., Garza, C. J., & Galaini-Wraight, S. (2000). Evaluation of the entomopathogenic fungi *Beauveria bassiana* and *Paecilomyces fumosoroseus* for microbial control of the silverleaf whitefly, *Bemisia argentifolii*. *Biological Control, 17*, 203–217.

Part 3

Environmental Risks Associated With Bees

9 Environmental Impact of Exotic Bees Introduced for Crop Pollination

Carlos H. Vergara

Introduction

Many bee species have been introduced for crop pollination around the world. The superfamily Apoidea (bees and sphecoid wasps) constitutes a very diverse group of insects, with more than 20,000 recognized species. The most important activity of bees, in terms of benefits to humans, is their pollination of all kinds of plants. Over 75% of the major world crops and 80% of all flowering plant species depend on insects for pollination (Nabhan & Buchmann, 1997). Bees are the main pollinators of more than 130 crop species in the United States and more than 400 crop species worldwide. They are the most important and common insect pollinators, accounting for more than 95% of the visits of insects to flowers. Bees pollinate 63 (77%) of the world's 82 most economically important plant species, and they are the most important known pollinator for 38 of those plant species (Delaplane & Mayer, 2000). The economic value of honey bee pollination has been estimated for the agricultural industry of several countries (table 9.1).

In the first section of this chapter, I consider the role of bees as pollinators and the motivation behind introductions and historical aspects of introductions. In later sections, I examine the information available on effects that introduced bees have on native fauna and flora and the effects of introduced pollinators on exotic fauna and flora for honey bees, bumble bees, megachilids, and halictids. In the penultimate section I consider some practical and legal aspects of introducing exotic bees, and in the final section I offer a conclusion on the subject of introducing bees as pollinators.

Table 9.1 Economic value of honey bee pollination for crop production.

Region	Value	Reference
United States	US $9 billion	Robinson et al. 1989
	US $1.6–5.7 billion	Southwick and Southwick 1992
Canada	Can $443 million	Scott-Dupree et al. 1995
United Kingdom	£202 million (all insect pollination)	Carreck and Williams 1998
	£137.8 million (honey bee pollination)	
European Union	US $4.3 billion	Borneck and Bricout 1984
		Borneck and Merle 1989
Australia	Aus $17 billion	Gordon and Davis 2003

Reasons for Introducing Bees Outside Native Ranges

Bees in general are regarded as beneficial insects for their role in pollination, and as a result, several species have been purposely introduced outside their natural ranges. However, the number of documented accidental introductions is far greater than the number of intended introductions. For example, of the 21 exotic bee species found in North America, 17 were accidental introductions (Cane, 2003). In a few cases, the introduced species were not introduced for pollination purposes.

Honey Bees

Honey bees (genus *Apis* L.) are originally from the Old World. Eleven *Apis* spp. are currently recognized (Michener, 2000), of which *Apis mellifera*, *A. cerana*, and *A. florea* have been domesticated, are used for beekeeping in different regions of the world, and have been introduced outside their natural ranges. *Apis mellifera* has diverged into more than 25 subspecies. Although native to Europe and Africa, many different races of *A. mellifera* have been introduced to virtually every country in the world, with the exception of Antarctica. Most of these introductions occurred during the colonization process by European settlers around the world. The Asiatic hive bee, *A. cerana*, is native to Asia between Afghanistan and Japan and from Russia and China in the north to southern Indonesia. Hives of this bee have been taken from Baluchistan to Iran (Crane, 1995), and this species has been introduced recently to Papua New Guinea (Bradbear & MacKay, 1995). *Apis florea* is native to Oman, spreading southeast through Asia as far as some of the islands of Indonesia. In recent years, it was introduced to Sudan and has been lately reported in Iraq (Glaiim, 2005).

Nine subspecies of *A. mellifera* were introduced into the United States before the 1920s. Most of these subspecies were from Europe (*A. m. mellifera*, *A. m. iberica*, *A. m. caucasica*, *A. m. ligustica*, *A. m. carnica*), but some were from Africa (*A. m. intermissa* and *A. m. lamarcki*) and the Middle East (*A. m. cypria* and *A. m. syriaca*). In the 1990s, the descendents of one more subspecies, the African *A. m. scutellata* (referred to as Africanized honey bees), arrived from Mexico and expanded in the southwestern United States (Pinto et al., 2003). Two subspecies were introduced by Spanish and Portuguese settlers into most of Latin America, *A. m. iberica* (the Spanish bee) and *A. m. mellifera* (the German bee).

The first known successful importation of Italian queen bees to the United States was made in 1860 (Pellett, 1938). At the beginning of the twentieth century, the Italian bee, *A. m. ligustica*, was introduced from the United States to many countries and regions in Latin America, including the Yucatan peninsula, where no previous introduction of *A. mellifera* had occurred.

In 1956 the Brazilian government commissioned Warwick Kerr to introduce African bees (*A. m. scutellata*) to Brazil to produce a new breed of honey bees. This new breed was to be less defensive than the wild African bees but more productive than European honey bees in Brazil's tropical setting. At the time, Brazil ranked 47th among the world's honey-producing countries. With the arrival of the new variety, that country's ranking quickly rose to seventh.

Honey bees (most likely *A. m. mellifera*) were introduced into Australia in 1810 by Samuel Marsden, who imported an unknown number of colonies from England. The early settlers attempted to use the honey bees for pollination of their fruit trees, but the first attempts to establish bee colonies failed. A second successful introduction was made in 1822, and further introductions to Australia were made over the next 50–60 years. Honey bees were probably not widespread until about 1930, although precise information on the rate of introduction and distribution is not available (Paton, 1996).

Reasons for Introducing Honey Bees Outside Native Ranges

Most of the initial introductions of honey bees to new regions were done to produce honey and wax or to improve the production that was obtained with the species or races previously used. Other bee products, such as pollen, bees (as packages or nuclei), and queens, have become of commercial importance more recently. Pollination as a commercial activity has become highly relevant in the past 50 to 60 years, except for the aforementioned case of early introductions into Australia.

Honey bees are generalists and visit a wide array of plants in bloom during one season. Being generalists, they are not the best pollinators for every crop and actually visit less than one-third of all plant species in a given locality (Butz-Huryn, 1997). However, honey bee colonies can be managed, transported and manipulated very efficiently for most crops. They can also be selected for increased pollen hoarding.

In the United States the majority of honey bee colonies rented by growers are used on only 13 crops, and 2–2.5 million out of a total 2.9 million colonies are rented for pollination each year (Morse & Calderone, 2000). In 2004 the almond crop alone required

1.4 million honey bee colonies. Given the growth in almond acreage projected for the next 6 years, California will require about 2 million colonies for almond pollination by 2012 (Sumner & Boriss, 2006). More than 47,000 colony rentals take place every year in Canada (Scott-Dupree et al., 1995), and in Mexico about 200,000 colonies were rented in 1999 for the pollination of at least 15 crops, including the major export crops (Lastra-Marín & Peralta-Arias, 2000). Australia has no nationwide figures on the number of colonies rented for pollination, but it is estimated that 40,000 honey bee colonies are used each year for crop pollination in the state of Victoria, which has about 20% of the total number of hives in Australia and is the most important horticultural area in that country. It appears that expansion of the almond industry in Victoria within 6 or 7 years will require more than 40,000 colonies for almonds alone. Some of these colonies may need to be sourced from New South Wales (Benecke, 2003).

Effects of Introduced Honey Bees on Native Plants

Honey bees could alter the pollination rates of plants in several different ways (see also chapter 10, this volume). They could: (1) add to the services provided by native pollinators and increase seed production; (2) displace native pollinators from flowers without providing equivalent pollination services, leading to declines in seed production; (3) alter the behavior of native pollinators in ways that alter patterns of pollen dispersal, leading to changes in seed production; and (4) remove pollen from flowers and thus reduce the quantities of pollen being transferred to flowers by legitimate pollinators, leading to a decrease in seed production (Paton, 1996).

Honey bees are major or secondary pollinators of many native plants in Australia, New Zealand, and the Americas. In these regions, honey bees visit a wide variety of plants but tend to intensively utilize 15–25% of the species available (reviewed in Butz-Huryn, 1997). Honey bees provide efficient pollination for some native plants in Australia, North America, and South America. They also have been shown to be floral parasites of native plants in Australia, North America, and Jamaica and to reduce seed set of native species in the neotropics and in Australia (Goulson, 2003).

The effects of honey bees on the production of seeds by native plants have been documented in different regions of the world. In Australia, seed production for some plants was reduced when honey bees were frequent floral visitors (e.g., *Callistemon rugulosus* [Myrtaceae]; Paton, 1993), whereas for others seed production was enhanced (e.g., *Banksia ornata* [Proteaceae]; Paton, 1996). Studies on bird-adapted Australian plants such as *Correa reflexa* (Rutaceae; Paton, 1993) and *Brachyloma ericoides* (Epacridaceae; Celebrezze & Paton, 2004) showed that native birds contributed significantly to fruit set even though honey bees were much more frequent visitors to these flowers. Bird exclusion experiments performed with the latter plant species produced a mean percentage of 12.3 ± 2% fruit set, whereas open pollinated plants produced 21 ± 2% fruit set. Plant species whose seed production increased were those that received inadequate attention from their native pollinators (Paton, 1996). Plant-pollinator systems are vulnerable to perturbations such as habitat clearance and degradation, and some Australian plants may now depend on honey bees for full pollination because their native pollinators have declined dramatically or even disappeared in some areas.

In Brazil, introduced honey bees were shown to reduce pollination success of *Clusia arrudae* (Clusiaceae), a dioecious species pollinated by *Eufriesea nigrohirta* (Apidae, Euglossini) that visit *C. arrudae* flowers to collect resin. Honey bees remove an excess of pollen without performing pollination. Male flowers, however, are also visited by individuals of *A. mellifera*, which remove about 99% of their pollen grains. When *E. nigrohirta* bees depart from flowers previously visited by *A. mellifera*, they carry on their bodies less than 0.1% of the pollen grains carried by *E. nigrohirta* bees that depart from flowers not visited by honey bees. This may explain why the frequency of *A. mellifera* at male flowers is negatively correlated with the number of seeds produced by female flowers. Pollen depletion by *A. mellifera*, therefore, reduces the effectiveness of the native pollinator (Carmo et al., 2004). In a study on pollen movement in *Impatiens capensis* (Balsaminaceae), Wilson and Thomson (1991) found that pollen-collecting *A. mellifera* removed almost twice as much pollen as nectar-collecting *Bombus* when visiting a virgin-male phase flower but deposited an order of magnitude less on stigmas of virgin-female phase flowers.

Effects of Introduced Honey Bees on Native Pollinators

Introduced honey bees could potentially compete for floral resources with native bees and cause reductions in survival, growth, or reproduction in these bees. Research into honey bee-native bee competition has focused mainly on indirect measurements such as floral resource overlap, visitation rates, or resource harvesting, and any negative interaction has been interpreted as competitive exclusion of wild bees by honey bees. Although this research can be valuable in indicating the potential for competition between honey bees and native bees, assessments of fecundity, survival, or population density are needed to determine whether the long-term survival of a native bee species is threatened. Paini (2004) reviewed 38 studies on this subject, most of which used indirect measurement of competition. Many of these studies were compromised by low replication, confounding factors, or poor interpretation of results. Based on the few studies that quantified the impact of honey bees on native bee survival, fecundity, or population densities, Paini (2004) was unable to make any definite conclusion regarding the real impact of honey bees on wild bees. Butz-Huryn (1997) similarly concluded that there was little evidence that the presence of honey bees has any impact on native bees, just as Paini et al. (2005) found for a native Australian *Megachile* (Megachilidae). Thomson (2004) experimentally tested the effects of honey bees by evaluating foraging behavior and reproductive success of a native eusocial bee, *Bombus occidentalis* Greene, in coastal California. *Bombus occidentalis* colonies located near experimentally introduced honey bee hives had lower mean rates of forager return and a lower ratio of foraging trips for pollen relative to nectar. Estimates of both male and female reproductive success of *B. occidentalis* also showed a reduction with greater proximity to introduced honey bee hives. Reproductive success correlated significantly with measures of colony foraging behavior, most strongly with the relative allocation of foraging effort to pollen collection.

Few studies have focused on the interactions between introduced honey bees and native bird pollinators, but honey bees have been shown to have substantial niche overlap with nectarivorous birds. For example, the presence of honey bees in Arizona has been found to deter foraging by hummingbirds (Schaffer et al., 1983). In Southern Australia, honey bees are now the most frequent floral visitors and often remove more than 80% of the floral resources being produced, even from plants that are pollinated largely by birds (Paton, 1993, 1996). Removal of floral resources results in native fauna being displaced from flowers, and in one case local population densities of New Holland honeyeaters (*Phylidonyris novaehollandiae*, Meliphagidae) were reduced (Paton, 1993). Honey bees largely displaced honeyeaters from flowers of *Callistemon rugulosus* without effecting comparable pollination, leading to reduced seed production (Paton, 1993). In another bird-pollinated plant, *Correa reflexa*, removal of pollen from flowers by honey bees reduced the amounts of pollen that birds (mostly honeyeaters) would subsequently deliver to flowers (Paton, 1993). In contrast, in the case of the winter-flowering *Banksia ornata*, introduced honey bees are effective pollinators and, in fact, provide much-needed pollination service because of a shortage of *Banksia*'s native pollinators, nectar-feeding birds, including honeyeaters (Paton, 1997). Ironically, honey bees may have contributed to this shortage by consuming substantial quantities of the limited summer and autumn floral resources, thereby reducing honeyeater populations.

No study to date has examined competition for floral resources between introduced honey bees and nectarivorous bats on flowers. This type of interaction could occur in regions in which agaves and columnar cacti are important elements of the vegetation. Several studies in this kind of habitat have shown that diurnal visitation by honey bees and other insects may either act synergistically with bat visitation to increase reproductive success of the plants or be unimportant when the plants depend mainly on bats to set seed (Molina-Freaner and Eguiarte, 2003; Rocha et al., 2005).

Another negative interaction that may occur between bats and honey bees is competition for nest sites. This aspect has been studied in Australia, where a wide variety of vertebrates, including bats, may use hollows in trees for roosting and nesting (reviewed in Paton, 1996). However, there is no strong evidence of competition between feral honey bees and bats nesting in hollows. Concern about competition for nest sites is based largely on recorded instances of honey bees occupying hollows that had previously been used by bats, parrots, or nightjars (Paton, 1996). These displacements, however, may involve only a small proportion of the population and have no significant effect on the population sizes of native fauna, particularly if other hollows are available for use.

Bumble Bees

The first attempt to introduce bumble bees outside their natural distribution ranges took place in 1875, when queens of at least two species from England were released in New Zealand. The introduced bees were established in New Zealand by 1885. In the early 1980s *B. ruderatus* was introduced from New Zealand to Chile (Arretz & McFarlane,

1986). Currently, it is established in most mesic regions of Chile and is the most abundant floral visitor found in temperate forests of southwestern Argentina (Morales & Aizen, 2006). *Bombus terrestris* has been used for agricultural crop pollination since the late 1980s, primarily in Europe, but ultimately in more than 15 countries. In many countries in which it has been imported, populations of *B. terrestris* have become naturalized and are rapidly expanding their ranges. *Bombus terrestris* is currently used for greenhouse crop pollination in many countries outside its native geographic range. *Bombus impatiens* is the only species cultivated in North America on a commercial scale. *Bombus occidentalis* was also reared for use in the states west of the Rocky Mountains. However, the recurrent heavy infestations of *B. occidentalis* colonies with the fungal pathogen *Nosema bombi* caused severe problems, so that the mass breeding of this species was discontinued (Thorp, 2003).

Bombus terrestris

Due to increasing commercial availability of artificially reared colonies, *B. terrestris* is currently used for greenhouse crop pollination in many countries outside its native geographic range. The native range of *B. terrestris* covers all of Europe, the coastal area of North Africa, and the Middle East, as well as the western part of Asia. Within these areas, it is represented by a number of subspecies that differ in a variety of behavioral traits, including floral color preferences, flower detection, and learning behavior (Chitka et al., 2004). This species has been introduced into New Zealand (Macfarlane & Gurr, 1995), Tasmania (Goulson, 2003), Brazil, Chile, Mexico, Japan (Thorp, 2003), and Uruguay (Freitas et al., 2003), among other countries. It was also apparently introduced into mainland Australia (New South Wales) but did not establish (Froggatt, 1912; Rayment, 1935). In 2003, one worker and one queen of *B. terrestris* were found on mainland Australia (Dollin, 2003). Worldwide sales of *B. terrestris* are estimated to be around 850,000 colonies per year (Velthuis & van Doorn, 2006).

Effects of Introduced *Bombus terrestris* on Native and Introduced Plants

Bombus terrestris is the only bumble bee species introduced to Australia and New Zealand that visits native plants. Studies on preferences for introduced plants by *B. terrestris* in Australia and New Zealand have not provided corroborating evidence. In New Zealand, *B. terrestris* visits more species of introduced plants than native plants (Hanley & Goulson, 2003). On the Australian island of Tasmania, no difference in attractiveness of introduced plants and native plants to this bumble bee was found in a garden at the intersection of an urban area and native vegetation (Hingston, 2005).

To date, no studies are available on the effects of exotic bumble bees on the reproduction of native plants. *Bombus terrestris* has the potential to disrupt pollinator services by robbing bird-pollinated flowers (e.g., *Epacris impressa* [Epacridaceae] in Tasmania; Hingston & McQuillan, 1998). By definition, nectar robbers obtain a reward by piercing floral tissues without contacting the anthers and stigma, thereby failing to effectively transfer pollen (Inouye, 1979). When the structure of the flower makes access to the nectaries impossible, *B. terrestris* (and some other bee species) use their mandibles to bite through the base of the corolla and act as floral parasites, removing nectar without pollinating the plant. The consequences of this behavior are not easy to predict. Robbers can reduce the amount of reward available, causing decreased visitation rates by pollinators (McDade & Kinsman, 1980) and a reduction in seed set (Roubik, 1982; Roubik et al., 1985; Irwin & Brody, 1999). Robbing can also damage floral tissues, interfering with or preventing seed production (Galen, 1983). Nectar robbers may pay some legitimate visits to the plants they otherwise rob (Inouye, 1979). Thus nectar robbing may have little influence on plant fecundity if nectar robbers also collect pollen and, in doing so, pollinate the plant or if other pollinators are present (Newton & Hill, 1983; Arizmendi et al., 1995; Morris, 1996; Stout et al., 2000). Some plants may actually benefit from the activity of nectar robbers by forcing legitimate foragers to make more long-distance flights, hence increasing genetic variability through outcrossing (Zimmerman & Cook, 1985). They also can force legitimate pollinators to visit more flowers per unit time, thereby increasing seed set (Heinrich & Raven, 1972).

The most important effect that *B. terrestris* could have on introduced plants would be invasive mutualisms with weed species. This topic is dealt with in detail in chapter 10 of this book.

Effects of Introduced Bombus terrestris *on Native Pollinators*

The only study to date that addresses this question was carried out in Tasmania by Hingston and McQuillan (1999). Two species of native *Chalicodoma* (Megachilidae) bees were apparently displaced by *B. terrestris* foragers at *Gompholobium huegelli* (Fabaceae) flowers. The study did not, however, measure the impact of this displacement on reproductive aspects of the native megachilids.

The studies of Paini and Roberts (2005) and Paini et al. (2005) on the effects of honey bees on native Australian bees indicate that the response of bees to the presence of introduced pollinators, such as *B. terrestris*, would depend on the species involved and on specific adaptive traits of the native species.

Massive movement of nonnative populations of *B. terrestris* as a consequence of commercial production of colonies for greenhouse crop pollination within the distributional range of the species has been conducted with no risk assessment. *Bombus terrestris* populations differ significantly from each other in their genetic makeup, as demonstrated by strong differences in coat color and behavioral traits. For example, Ings et al. (2005) have shown that Canary Island bees (*B. t. canariensis*) were superior in their nectar foraging performance to Sardinian bees (*B. t. sassaricus*), which were generally superior to mainland European bumble bees (*B. t. terrestris*). These interpopulation differences in performance are largely explained by interpopulation variation in forager size, with larger bees being superior foragers. However, even when body size was accounted for, "native" bees were found not to be superior to transplanted non-native bees in all but one case. In conclusion, non-native populations, especially those with large individuals, can be highly competitive foragers. This could lead to the establishment of non-native bees and the displacement of native ones.

Effects of Introduced Bombus terrestris on Introduced Pollinators

No information is available on the possible effects of *B. terrestris* on foraging or reproduction of other introduced pollinators, such as *A. mellifera*. In New Zealand, where four different introduced bumble bee species occur, *B. terrestris* is the most abundant, whereas two of the other three species (*B. ruderatus* and *B. subterraneus*) are rare, and their populations are apparently declining. The fourth species, *B. hortorum*, is relatively abundant. However, these differences in abundance seem to be related to the abundance of introduced forage plants rather than competition between bumble bee species (Goulson & Hanley, 2004).

Bombus ruderatus

Bombus ruderatus was introduced to New Zealand between 1875 and 1906 and became established in both the North and the South Islands. Populations of *B. ruderatus* are now declining in New Zealand, as flower-rich meadows are being lost in that country (Goulson & Hanley, 2004). A possibility not examined by Goulson and Hanley (2004) is the occurrence of competitive displacement of the rare bumble bee species by *B. terrestris*.

This bumble bee was introduced to Southern Chile in 1982 from a New Zealand stock (Arretz & Macfarlane, 1986). Since its introduction near Malleco, it has spread toward the north to Chillán and to the south to Puerto Montt, and within 10 years of its introduction has spread into the temperate southern forests of Argentina (Abrahamovich et al., 2001; Morales & Aizen, 2002; Ruz, 2002).

Effects of Introduced Bombus ruderatus on Native and Introduced Plants

Bombus ruderatus forages exclusively on introduced plants in New Zealand (Goulson & Hanley, 2004) and prefers to visit introduced plants in Chile (Rebolledo et al., 2004). In temperate forests of northwest Patagonia (Argentina), *B. ruderatus* equally visited flowers of exotic and native plants in both disturbed and undisturbed habitats (Morales & Aizen, 2002). Red clover, *Trifolium pratense* (Fabaceae), depends heavily on *B. ruderatus* to set seed; for this reason *B. ruderatus* was introduced into New Zealand and Chile, where red clover seed production is of economic importance (Ruz, 2002).

Bombus impatiens

This species has a wide distribution east of the Rocky Mountains of the United States. *Bombus impatiens* is also used in Mexico, where it is not native. To date, the total yearly sales of B. impatiens to Canada, the United States, and Mexico combined is 70,000 colonies (Velthuis & van Doorn, 2006). The recent accelerated growth of greenhouse crops in Mexico has triggered the need for various elements of commercial production. One of these elements is the presence of bumble bee colonies for pollination. In 1997, Koppert de Mexico, a branch of the original Dutch company Koppert Biological Systems, was established in Mexico. Since 2001, Koppert de Mexico imports B. impatiens queens from Michigan and finishes colony buildup in Mexico. Koppert opened a bumble bee-rearing facility in Querétaro, north of Mexico City, in 2004. Currently this company claims to have 83% of the bumble bee domestic market, and it sends B. impatiens colonies from Querétaro to Sinaloa, Sonora, Baja California, Jalisco, Michoacán, and Querétaro.

Possible Effects of Introduced Bombus impatiens on Native Bumble Bees

Bombus impatiens is native to the eastern part of the United States, east of the Rocky Mountains. A narrow region in southwestern Texas separates its native range from those of its closest relatives, the Mexican B. ephippiatus and B. wilmattae. These three species belong to the subgenus *Pyrobombus*, which is clearly monophyletic (Cameron et al., 2007). Bombus ephippiatus and B. wilmattae are considered parts of a single variable species (Williams, 2005). The possibility of interbreeding between B. impatiens and B. ephippiatus has been confirmed by experimental crosses at the Bee Lab of the University of Guadalajara in Autlán (Cuadriello, personal communication). If imported B. impatiens can interbreed with wild populations of B. ephippiatus and B. wilmattae, local adaptations in these species could be reduced and their specific distinction lost. Their close relationship may also increase the susceptibility of the Mexican species to introduced parasites and diseases. Recently, a queen of B. impatiens was found in the wild near Ciudad Guzmán, Jalisco. This is a very clear indication that the species has escaped and become naturalized in Mexico, posing a possible threat to native bumble bees.

Because B. ephippiatus is so widespread and biologically similar to B. impatiens, it would make a good species for commercial rearing within its native range in Mexico. Bombus ephippiatus has been successfully reared in the laboratory and could be developed for commercial pollination.

The Mexican Federal Law of Animal Health of 1993 established monitoring, prevention, control, and eradication of diseases for all terrestrial animals in Mexico (Congreso de los Estados Unidos Mexicanos, 2004). The National Service for Agro-Food Health, Safety, and Quality (SENASICA) regulates the importation of agricultural products into Mexico, including requirements for three species of bumble bee: B. impatiens, B. occidentalis, and B. ephippiattus. The latter species is included in these regulations because in the 1990s the idea was discussed of exporting queens of this species to Europe to rear colonies and then reimporting them into Mexico. The restrictions require that official health certificates stating that the imported bumble bees are free of infections from *Nosema*, *Varroa*, and fungal diseases accompany each shipment into Mexico (SENASICA, 2006). Incoming colonies are then sent to the Animal Health Experimental

Service Center for examination for ectoparasites and *Nosema* infection, with documentation complying with the requirements of Article 24 of the federal law (SENASICA, 2006).

Megachilidae

Several species of this family have been introduced outside their natural range, both intentionally and accidentally. The best documented case is the alfalfa leafcutting bee (*Megachile rotundata*), which is a native of Eurasia. It was introduced by accident into North America and by intention for alfalfa pollination in South America and Australia, with varying degrees of success (see also chapter 7, this volume). American strains of this species have been reintroduced into Europe in an attempt to restore native populations. The *Centaurea* leafcutting bee (*Megachile apicalis*) was also accidentally introduced into North America from Eurasia. Several *Osmia* species have also been introduced from Asia into North America for pollination of almond, apples and other fruit trees.

Megachile rotundata

Megachile rotundata is endemic to Eurasia, extending from western Europe and North Africa through northern Iran and southern Siberia to Mongolia. Apparently, *M. rotundata* was accidentally introduced on the east coast of the United States, where it was first found in the 1930s, and its presence in this country was confirmed in 1947 (Stephen, 2003). It was intentionally introduced into Canada, first in 1962 (Hobbs, 1964), with reintroductions in 1964 and 1966 (Stephen, 2003). A first unsuccessful introduction of *M. rotundata* from the United States into Chile was done in 1963, followed by a massive introduction from the United States in 1971. These introductions were aimed at improving seed production of alfalfa (Ruz, 2002). This bee was introduced to Argentina in the 1970s (Martínez, 2001), to New Zealand in 1971 (Donovan, 1975) from Idahoan and Canadian stock, and from there to Australia in the early 1970s. Other introductions to southern Australia took place in 1970 (from Canada) and between 1988 and 1996 from New Zealand (Woodward, 1996). Between 1998 and 2005, variable numbers of *M. rotundata* have been imported to Australia (Anderson, 2006). Little information is available on populations of the introduced Megachilidae, but one study suggests that these solitary species do not attain high densities in Australia (Woodward, 1996).

Megachile rotundata is an oligolege that has a strong preference for alfalfa flowers, which they visit even when other plants are in bloom nearby, reaching very high pollination rates (80–100% in some cases; Bosch & Kemp, 2005). Commercially managed on a large scale, this species nests and completes its reproductive cycle within the alfalfa fields it pollinates, provided there are an adequate number of available nesting holes. High numbers of foraging *M. rotundata* females discouraged foraging by the native alkali bee, *Nomia melanderi* (Halictidae), in alfalfa fields in experimental trials in several localities in the state of Washington (Mayer & Johansen, 2003). Whether or not this affects the reproduction of *N. melanderi* was not answered by the study.

Megachile apicalis

Megachile apicalis is also endemic to Eurasia, extending from Western Europe north to Ukraine and east to Uzbekistan. Its introduction to the United States was accidental (Stephen, 2003). *Megachile apicalis* is an oligolege on *Centaurea* spp. (Astereaceae) and is sympatric with *M. rotundata* in California. *Megachile apicalis* emerges before *M. rotundata*, but at a time when its host plant is not yet in bloom. When *Centaurea* blooms, *M. apicalis* appears in great numbers and displaces *M. rotundata* from the nest sites. This aggressive displacement is considered a severe problem for farmers who use *M. rotundata* for alfalfa seed production (Stephen, 2003). Invasive mutualism occurs when two or more species facilitate each other in establishment and spread into a new geographic region (Richardson et al., 2000). Bees and weeds can be invasive mutualists, such as is the case of honey bees and the yellow star-thistle (Barthell et al., 2001). Invasive mutualism may also occur between other introduced thistle species (*Centaurea* sp.) and *M. apicalis* (Barthell et al., 2003). Also, *M. apicalis* displays an aggressive nesting behavior that has the potential to affect other cavity-nesting bees, but this possible interaction has not been documented (Barthell et al., 2003).

Osmia cornifrons

Osmia cornifrons, the horn-faced bee, is a native to Japan and was first imported for apple pollination from Japan into Utah in 1965, but it did not survive there. In 1976, it was imported into Maryland, where it thrived in a climate that resembles that of central Japan (Batra, 1979). These bees strongly prefer the flowers of Rosaceae, and, unlike honey bees, they are not readily distracted by dandelions and other weeds. An individual horn-faced bee is 80 times more effective than an individual worker honey bee for pollination of apples (Maeta, 1990).

No information is available on possible effects of *O. cornifrons* on native plants or on other pollinators.

Halictidae

Nomia melanderi, the alkali bee, is endemic to western North America and is used in the United States for alfalfa pollination. It was introduced to New Zealand in 1971 for alfalfa seed production (Donovan, 1975, 1979). Alkali bee females collect pollen from composite or alfalfa flowers when they are abundant near nest sites. In New Zealand, alkali bees have not been seen visiting native flowers, although pollen removed from the scopae of a female alkali bee at a coastal nest site included 1% of a native *Selliera* sp. (Goodeniaceae; Donovan, 1980).

The number of nests, and thus the number of female alkali bees, increased from 70 in 1971 to about 14,000 in 1980. Large populations will develop only in the very few localized suitable nest soils or in the vicinity of alfalfa seed fields where constructed nest sites are provided (Donovan, 1980).

Practical and Legal Aspects of Introducing Exotic Bees

Most introductions of bees have been carried out without prior assessment of the potential impacts of these organisms on the environment. Almost all laws and regulations affecting the importation of bees focus on preventing diseases and parasites associated with plants and honey bees without considering the potentially adverse environmental impacts associated with the bees themselves (Flanders et al., 2003).

The risks to wild populations that have occurred or that could be caused by exotic pollinators have led several governments to impose restrictions on the importation of wild bees. The Wildlife and Countryside Act (1981) of the United Kingdom, for example, prohibits the release of any non-native animal into the environment, including in semiconfined situations such as in commercial greenhouses. The same kinds of concern made the government of the Canary Islands demand that only *B. canariensis* be used in the greenhouses in their territory. The Japanese government has included *B. terrestris* in its Invasive Alien Species Act (Velthuis & van Doorn, 2006), and future importation will likely be banned. China and South Africa do not allow *B. terrestris* to be imported (Velthuis & van Doorn, 2006). In New South Wales, the alteration of natural pollination dynamics caused by the presence of *B. terrestris* has prompted its listing as a key threatening process (Department of Environment and Conservation of New South Wales, 2004), and in Victoria, Australia, it is listed as a potentially threatening process (Victorian Scientific Advisory Committee, 2000).

In North America, the United States prohibits or restricts the importation of honey bees to prevent the introduction of pests and parasites. The importation of non-*Apis* pollinators is regulated under the Plant Protection Act of 2000, with the main focus on preventing the introduction of parasites and pathogens. The Canadian Food Inspection Agency (CFIA) is responsible for regulating bees under the 1990 Health of Animals Act, and importations are regulated under the Honeybee Prohibition Regulations of 2004. Similar to the United States, the Canada Plant Protection Act of 1990 covers other pollinating species by administration from the Plant Products Directorate, Plant Health Division, Export/Import Section. As mentioned before, in Mexico the Federal Law of Animal Health of 1993 established monitoring, prevention, control, and eradication of diseases for all terrestrial animals in Mexico (*Ley Federal de Sanidad Animal*, 2004). The importation of agricultural products into Mexico, including three species of bumble bees, is regulated by the National Service for Agro-Food Health, Safety, and Quality, requiring official certificates of bee health.

The complex balance among costs, benefits, and stakeholders involved in making decisions about whether or not to introduce pollinators is well exemplified by the case of *Anthophora plumipes* (Batra, 2003). This species was brought from Japan between 1989 and 1995, and the bees were reared and kept at the U.S. Department of Agriculture

(USDA) Beltsville Agricultural Research Center (BARC). Complaints about this bee, stating possible interference with native bees, were sent to the USDA Animal and Plant Health Inspection Service (APHIS). On account of the complaints, APHIS decided to order the destruction of the bees. Half of the population was destroyed and the other half was kept at BARC. Subsequent studies demonstrated that *A. plumipes* did not migrate farther than 30 m from their native nests (Batra, 2003). However, recent collecting reveals that *A. plumipes* is well established in Maryland and vicinity (Ascher, 2006).

Criteria for selecting candidate pollinators are available in the literature (e.g., Donovan, 1990). These lists are prepared with specific crops in mind, most of which are also species exotic to the country that issues the regulations. Thus the recommended pollinator is often an exotic species. This circular kind of reasoning has hindered the development of native pollinators, especially in countries in which research on this topic is incipient or nonexistent. An example that clearly illustrates this point is the use of imported bumble bees for pollination in countries in which native species could be used for the same purposes. It is technically possible to breed colonies locally, in large numbers and at times of the year when they are needed. This type of breeding, however, would have a seasonal character, and starting the production anew takes much more effort than its continuation. Furthermore, because the dearth period for selling also needs to be compensated, locally produced colonies would become more expensive than those that come from a year-round regular producer. Apparently, this economic difficulty has prevented the emergence of local production of colonies in several countries, with the consequence that the potential risks of importing colonies have to be accepted. In this case, the efficient methods of rearing the imported bumble bees are a compounding factor against the development of local pollinators. In any country, the local tomatoes can most probably be pollinated by those bees that prefer the native Solanaceae; if there is concern about the dangers inherent in importation, the government, as well as the farmers, should invest in research if they want to compete in the international market (e.g., Hogendoorn et al., 2000; Estay et al., 2001). From the viewpoint of safety and nature protection, however, production of a native pollinating species should be encouraged.

Conclusions

The introduction of bees for crop pollination (or for other purposes) has the potential to affect plants and pollinators, both native and introduced. One effect of exotic introduction is the disruption of an economic activity at national and regional levels, as has occurred with the honey beekeeping industry in most countries in the Americas after the introduction of African honey bees to Brazil. Exotic bees can also disrupt the reproduction of native and introduced plants; they sometimes have negative impacts on native bee survival, fecundity, population densities, foraging behavior, and reproductive success.

Providing scientific evidence of these changes is not easy. However, a few recent studies demonstrate that exotic pollinators affect reproductive output of native species of pollinators. Evidence of disruption of plant reproduction is more readily available. Invasive mutualisms between introduced pollinators and introduced weeds also have been demonstrated (see chapter 10, this volume).

In my opinion, the use of exotic pollinators should be considered as a last resort, and preference should be given to the development of native pollinators over importation of pollinators developed elsewhere. However, in some cases a compromise has to be reached between the need for pollinators for crops and the commercial feasibility of developing native species as pollinators.

The impacts of introducing bees are sometimes unanticipated, as with the introduction of other beneficial species such as biocontrol agents (reviewed in Richardson et al., 2000). In some cases, the target hosts cannot support the population of the introduced agent in the long term, and nontarget species may suffer attack during a transient period of spillover parasitism shortly after the introduction of the agent. Even though this period may not be long, it can cause extinctions (Lynch et al., 2002). A similar situation could occur with the introduction of pollinating species. Bohart (1962) pointed out three dangers inherent in the introduction of foreign pollinators to the United States: (1) the introduction of unwanted arthropods or other organisms, (2) possible damage to plants, and (3) the competitive displacement of local species.

A case that illustrates the process of accepting or rejecting the importation of exotic bees is currently unfolding in Australia regarding the importation of *B. terrestris* for greenhouse tomato pollination. Tomato growers, represented by the Australian Hydroponic and Greenhouse Association (AHGA), advocate the use of this insect as the only realistic alternative for economically efficient pollination of tomatoes. Meanwhile, scientists and interested citizens, led by the Australian Native Bee Research Centre, oppose this importation, arguing potential threats to agriculture, native bees, birds, plants, and urban gardens. At the same time, a group of researchers offered a native bee as an alternative to bumble bees (Hogendoorn et al., 2006).

Whatever the outcome of this case, the process of presenting evidence from both sides is actively under way, and the regulatory agency responsible for the decision will have solid information on which to base its decisions. Most importations done in the past and in other parts of the world were done without considering any background information, and the effects of such importations are now irreversible.

Extirpation of undesirable organisms is impossible or impractical in most cases, but the lessons from the past need to be taken into account when planning modifications to the environment, as in the case of proposing the introduction of alien species. The effect of introduced pollinators on native ecosystems could appear on a scale that seems unimportant or could go unnoticed for long periods of time. Accepting to further modify the environment knowingly is not only irresponsible but is also unethical.

References

Abrahamovich, A. H., Tellería, M. C., & Díaz, N. B. (2001). *Bombus* species and their associated flora in Argentina. *Bee World, 82*, 76–87.

Anderson, D. (2006). Improving lucerne pollination with leafcutter bees: Stage 2 (Publication No. 06/108, Project No. CSE-91). Canberra, Australia: Rural Industries Research and Development Corporation.

Arizmendi, M. C., Domínguez, C. A., & Dirzo, R. (1995). The role of an avian nectar robber and of hummingbird pollinators in the reproduction of two plant species. *Functional Ecology, 10,* 119–127.

Arretz, P. V., & Macfarlane, R. P. (1986). The introduction of B. *ruderatus* to Chile for red clover pollination. *Bee World, 67,* 15–22.

Ascher, J. (2006). Introducing pollinators to new areas? Message posted on March 7, 2006, to http://lists.sonic.net/pipermail/pollinator/2006-March/000219.html.

Barthell, J. F., Randall, J. M., Thorp, R. W., & Wenner, A. M. (2001). Promotion of seed set in yellow star-thistle by honey bees: Evidence of an invasive mutualism. *Ecological Applications, 11,* 1870–1883.

Barthell, J. F., Thorp, R. W., Frankie, G. W., Kim, J. Y., & Hranitz, J. M. (2003). Impacts of introduced solitary bees on natural and agricultural systems: The case of the leafcutting bee, *Megachile apicalis* (Hymenoptera: Megachilidae). In K. Strickler & J. H. Cane (Eds.), *For nonnative crops, whence pollinators of the future?* (151–162). Lanham, MD: Entomological Society of America.

Batra, S. W. T. (1979). *Osmia cornifrons* and *Pithitis smaragdula,* two Asian bees introduced into the U.S. for crop pollination. In *Proceedings of the IV International Symposium on Pollination* (207–312). Colleg Park, MD: Maryland Agricultural Experimental Station.

———. (2003). Bee introductions to pollinate our crops. In K. Strickler & J. H. Cane (Eds.), *For nonnative crops, whence pollinators of the future?* (85–98). Lanham, MD: Entomological Society of America.

Benecke, F. (2003). *Commercial beekeeping in Australia* (Publication No. 03/037, Project No FSB-1A). Canberra, Asutralia: Rural Industries Research and Development Corporation.

Bohart, G. E. (1962). Introduction of foreign pollinators, prospects and problems. In *Proceedings of the First International Symposium on Pollination, Copenhagen* (Communication No. 7, 181–188). Svalöv, Sweden : Swedish Seed Growers' Association.

Borneck, R., & Bricout, J. P. (1984). Evaluation de l'incidence économique de l'entomofaune pollinisatrice en agriculture. *Bulletin of Technical Apiculture, 11,* 47, 117–124.

Borneck, R., & Merle, B. (1989). Essai d'une evaluation de l'incidence economique de l'abeille pollinisatrice dans l'agriculture européenne. *Apiacta, 24,* 33–38.

Bosch, J., & Kemp, W. P. (2005). Alfalfa leafcutting bee population dynamics, flower availability, and pollination rates in two Oregon alfalfa fields. *Journal of Economic Entomology, 98,* 1077–1086.

Bradbear, N., & MacKay, K. (1995). Developing agencies funding and interinstitutional cooperation. In P. Kevan (Ed.), *The Asiatic honey bee: Apiculture, biology, and role in sustainable development in tropical and subtropical Asia* (271–292). Cambridge, Ontario, Canada: Enviroquest.

Butz-Huryn, V. M. (1997). Ecological impacts of introduced honey bees. *Quarterly Review of Biology, 72,* 275–297.

Cameron, S. A., Hines, H. M., & Williams, P. H. (2007). A comprehensive phylogeny of the bumble bees (*Bombus*). *Biological Journal of the Linnean Society, 91,* 161–188.

Cane, J. H. (2003). Exotic nonsocial bees (Hymenoptera: Apiformes) in North America: Ecological implications. In K. Strickler & J. H. Cane (Eds.), *For nonnative crops, whence pollinators of the future?* (113–126). Lanham, MD: Entomological Society of America.

Carmo, R. S., Franceschinelli, E. V., & Silveira, F. A. (2004). Introduced honey bees (*Apis mellifera*) reduce pollination success without affecting the floral resource taken by native pollinators. *Biotropica, 36,* 371–376.

Carreck, N., & Williams, I. (1998). The economic value of bees in the UK. *Bee World, 79,* 115–123.

Celebrezze, T., & Paton, D. C. (2004). Do introduced honey bees (*Apis mellifera,* Hymenoptera) provide full pollination service to bird-adapted Australian plants with small flowers? An experimental study of *Brachyloma ericoides* (Epacridaceae). *Austral Ecology, 29,* 129–136.

Chitka, L., Ings, T. C., & Raine, N. E. (2004). Chance and adaptation in the evolution of island bumblebee behaviour. *Population Ecology, 46,* 243–251.

Congreso de los Estados Unidos Mexicanos (2004). Ley Federal de Sanidad Animal. Diario Oficial de la Federación. México, D. F. México. Retrieved February 25, 2008, from http://vlex.com.mx/vid/28095064.

Crane, E. (1995). History of beekeeping with *Apis cerana* in Asia. In P. Kevan (Ed.), *The Asiatic honey bee: Apiculture, biology, and role in sustainable development in tropical and subtropical Asia* (3–18). Cambridge, Ontario, Canada: Enviroquest.

Delaplane, K. S., & Mayer, D. F. (2000). *Crop pollination by bees.* Wallingford, Oxon, UK: CABI.

Department of Environment and Conservation of New South Wales. (2004). Introduction of the large earth bumblebee, *Bombus terrestris:* Key threatening process declaration. Retrieved February 15, 2007, from http://www.nationalparks.nsw.gov.au/npws.nsf/content/bombus terrestris ktp declaration.

Dollin, A. (2003). Bumblebees buzzing. *Feral Herald: Newsletter of the Invasive Species Council, 1*(4), 1–2.

Donovan, B. J. (1975). Introduction of new bee species for pollinating lucerne. *Proceedings of the New Zealand Grasslands Association, 36,* 123–128.

———. (1979). Importation, establishment and propagation of the alkali bee *Nomia melanderi* Cockerell (Hymenoptera: Halictidae) in New Zealand. In *Proceedings of the 4th International Symposium on Pollination* (257–268). College Park, MD: Maryland Agricultural Experimental Station.

———. (1980). Interactions between native and introduced bees in New Zealand. *New Zealand Journal of Ecology, 3,* 104–116.

———. (1990). Selection and importation of new pollinators to New Zealand. *New Zealand Entomologist, 13,* 26–32.

Estay, P., Wagner, A, & Escaff, M. (2001). Evaluación de *Bombus dahlbomii* (Guér.) como agente polinizador de flores de tomate (*Lycopersicon esculentum* (Mill.), bajo condiciones de invernadero. *Agricultura Técnica (Chile), 61,* 113–119.

Flanders, R. V., Wehling, W. F., & Craghead, A. L. (2003). Laws and regulations on the import, movement, and release of bees in the United States. In K. Strickler & J. H. Cane (Eds.), *For non-native crops, whence pollinators of the future?* (99–111). Lanham, MD: Entomological Society of America.

Freitas, B. M., Martins, C., Wittmann, D., Santos, I. A., Cane, J., Ribeiro, M., (2003). Bee management for pollination purposes: Bumblebees and solitary bees. Report of activities and preliminary results: São Paulo Declaration on Pollinators Plus 5 Forum, October, São Paulo, SP, Brazil.

Froggatt, W. W. (1912). Suggested importation of humble bees. *Agricultural Gazette, 23* (12), 896.

Galen, C. (1983). The effects of nectar thieving ants on seed set in floral scent morphs of *Polemonium viscosum. Oikos, 41*, 245–249.

Glaiim, M. K. (2005). *First definitive record of* Apis florea *in Iraq.* Retrieved March 12, 2006, from http://www.beesfordevelopment.org/info/info/species/first-definitive-record-o.shtml.

Gordon, J., & Davis, L. (2003). Valuing honey bee pollination (Publication No. 03/077, Project No. CIE-15A). Canberra, Australia: Rural Industries Research and Development Corporation.

Goulson, D. (2003). Effects of introduced bees on native ecosystems. *Annual Review of Ecology, Evolution, and Systematics, 34*, 1–26.

Goulson, D., & Hanley, M. E. (2004). Distribution and forage use of exotic bumblebees in South Island, New Zealand. *New Zealand Journal of Ecology, 28*(2), 225–232.

Hanley, M. E., & Goulson, D. (2003). Introduced weeds pollinated by introduced bees: Cause or effect? *Weed Biology and Management, 3*, 204–212.

Heinrich, B., & Raven, P. H. (1972). Energetics and pollination ecology. *Science, 176*, 597–602.

Hingston, A. B. (2005). Does the introduced bumble bee, *Bombus terrestris* (Apidae), prefer flowers of introduced or native plants in Australia? *Australian Journal of Zoology, 53*, 29–34.

Hingston, A. B., & McQuillan, P. B. (1998). Nectar robbing in *Epacris impressa* (Epacridaceae) by the recently introduced bumble bee *Bombus terrestris* (Apidae) in Tasmania. *Victorian Naturalist, 115*, 116–119.

———. (1999). Displacement of Tasmanian native megachilid bees by the recently introduced bumble bee *Bombus terrestris* (Linnaeus, 1758) (Hymenoptera: Apidae). *Australian Journal of Ecology, 47*, 59–65.

Hobbs, G. A. (1964). Importing and managing the alfalfa leaf-cutter bees (Canada Department of Agriculture Publication No. 1209). Ottawa, Ontario: Canada Department of Agriculture.

Hogendoorn, K., Gross, C. L., Sedgley, M., & Keller, M. A. (2006). Increased tomato yield through pollination by native Australian *Amegilla chlorocyanea* (Hymenoptera: Anthophoridae). *Journal of Economic Entomology, 99*(3), 828–833.

Hogendoorn, K., Steen, Z., & Schwarz, M. P. (2000). Native Australian carpenter bees as a potential alternative to introducing bumblebees for tomato pollination in greenhouses. *Journal of Apicultural Research, 39*, 67–74.

Ings, T. C., Schikora, J., & Chittka, L. (2005). Bumblebees: Humble pollinators or assiduous invaders? A population comparison of foraging performance in *B. terrestris. Oecologia, 144*, 508–516.

Inouye, D. W. (1979). The terminology of floral larceny. *Ecology, 61*, 1251–1253.

Irwin, R. E., & Brody, A. K. (1999). Nectar-robbing bumble bees reduce the fitness of *Ipomopsis aggregata* (Polemoniaceae). *Ecology, 80*, 1703–1712.

Lastra-Marín, I. J., & Peralta-Arias, M. A. (2000). *Situación actual y perspectiva de la Apicultura en México 2000.* México, D. F. Mexico: Secretaría de Agricultura y Ganadería.

Lynch, L. D., Ives, A. R., Waage, J. K., Hochberg, M. E., & Thomas, M. B. (2002). The risks of biocontrol: Transient impacts and minimum nontarget densities. *Ecological Applications 12*, 1872–1882.

Macfarlane, R. P., & Gurr, L. (1995). Distribution of bumble bees in New Zealand. *New Zealand Entomology, 18*, 29–36.

Maeta, Y. (1990). Utilization of wild bees. *Farming Japan, 24*, 13–19.

Martínez, E. (2001). *Polinización de alfalfa, Megachile rotundata* (Hoja Informativa 14). Buenos Aires, Argentina: Instituto Nacional de Tecnología Agropecuaria.

Mayer, D. F., & Johansen, C. A. (2003). The rise and decline of *Nomia melanderi* (Hymenoptera: Halictidae) as a commercial pollinator for alfalfa seed. In K. Strickler & J. H. Cane (Eds.), *For nonnative crops, whence pollinators of the future?* (139–150). Lanham, MD: Entomological Society of America.

McDade, L. A., & Kinsman, S. (1980). The impact of floral parasitism in two neotropical hummingbird-pollinated plant species. *Evolution, 34*, 944–958.

Michener, C. D. (2000). *The bees of the world*. Baltimore, MD: Johns Hopkins University Press.

Molina-Freaner, F., & Eguiarte, L. E. (2003). The pollination biology of two paniculate agaves (Agavaceae) from northwestern Mexico: Contrasting roles of bats as pollinators. *American Journal of Botany, 90*, 1016–1024.

Morales, C. L., & Aizen, M. A. (2002). Does invasion of exotic plants promote invasion of exotic flower visitors? A case study from the temperate forests of southern Andes. *Biological Invasions, 4*, 87–100.

Morales, C. L., & Aizen, M. A. (2006). Invasive mutualisms and the structure of plant–pollinator interactions in the temperate forests of north-west Patagonia, Argentina. *Journal of Ecology, 94*, 171–180.

Morse, R. A., & Calderone, N. W. (2000). The value of honey bees as pollinators of U.S. crops in 2000. *Bee Culture, 132*, 1–19.

Morris, W. F. (1996). Mutualism denied: Nectar robbing bumble bees do not reduce female or male success of bluebells. *Ecology, 77*, 1451–1462.

Nabhan, G. P., & Buchmann, S. L. (1997). Services provided by pollinators. In G. Daily (Ed.), *Nature's services: Societal dependence on natural ecosystems* (133–150). Washington, DC: Island Press.

Newton, S. D., & Hill, G. D. (1983). Robbing of field bean flowers by the short-tongued bumble bee *Bombus terrestris* L. *Journal of Apicultural Research, 22*, 124–129.

Paini, D. R. (2004). The impact of the introduced honey bee (*Apis mellifera*) (Hymenoptera: Apidae) on native bees: A review. *Austral Ecology, 29*, 399–407.

Paini, D. R., & Roberts, J. D. (2005). Commercial honey bees (*Apis mellifera*) reduce the fecundity of an Australian native bee (*Hylaeus alcyoneus*). *Biological Conservation, 123*, 103–112.

Paini, D. R., Williams, M. R., & Roberts, J. D. (2005). No short-term impact of honey bees on the reproductive success of an Australian native bee. *Apidologie, 36*, 613–621.

Paton, D. C. (1993). Honey bees in the Australian environment. Does *Apis mellifera* disrupt or benefit the native biota? *Bioscience, 43*, 95–103.

———. (1996). *Overview of the feral and managed honey bees of Australia: Distribution, abundance, extent of interactions with native biota, evidence of impacts and future research*. Canberra: Australian Nature Conservation Society.

———. (1997). Honey bees and the disruption of plant-pollinator systems in Australia. *Victorian Naturalist, 114*, 23–29.

Pellett, F. C. (1938). *History of American beekeeping*. Ames, IA: Collegiate Press.

Pinto, M. A., Johnston, J. S., Rubink, W. L., Coulson, R. N., Patton, J. C., & Sheppard, W. S. (2003). Identification of Africanized honey bee (Hymenoptera: Apidae) mitochondrial DNA: Validation of a rapid polymerase chain reaction-based assay. *Annals of the Entomological Society of America, 96*, 679–684.

Rayment, T. (1935). *A cluster of bees*. Sydney, Australia: Endeavour Press.

Rebolledo, R. R., Martínez, H., Palma, R., Aguilera, A., & Klein, C. (2004). Actividad de visita de *B. dahlbomi* (Guérin) y *B. ruderatus* (F.) (Hymenoptera: Apidae) sobre trébol rosado (*Trifolium pratense* L.) en la IX región de la Araucanía, Chile. *Agricultura Técnica (Chile), 64*, 245–250.

Richardson, D. M., Pysek, P., Rejmánek, M., Barbour, M. G., Panneta, F. D., & West, C. J. (2000). Naturalization and invasion of alien plants: Concepts and definitions. *Diversity and Distributions, 6*, 93–107.

Robinson, W. S., Nowodgrodzki, R., & Morse, R. A. (1989). The value of bees as pollinators of U.S. crops. *American Bee Journal, 129*, 411–423, 477–487.

Rocha, M., Valera, A., & Eguiarte, L. E. (2005). Reproductive ecology of five sympatric *Agave littaea* (Agavaceae) species in central Mexico. *American Journal of Botany, 92*, 1330–1341.

Roubik, D. W. (1982). The ecological impact of nectar robbing bees and pollinating hummingbirds on a tropical shrub. *Ecology, 63*, 354–360.

Roubik, D. W., Holbrook, N. M., & Parra, G. (1985). Roles of nectar robbers in reproduction of the tropical treelet *Quassia amara* (Simaroubaceae). *Oecologia, 66*, 161–167.

Ruz, L. (2002). Bee pollinators introduced to Chile: A review. In P. Kevan & V. L. Imperatriz Fonseca (Eds.), *Pollinating bees: The conservation link between agriculture and nature* (155–167). Brasilia, Brazil: Ministry of Environment.

Schaffer, W. M., Zeh, D. W., Buchmann, S. L., Kleinhans, S., Valentine Schaffer, M., & Antrim, J. (1983). Competition for nectar between introduced honey bees and native North American bees and ants. *Ecology, 64*, 564–577.

Scott-Dupree, C., Winston, M., Hergert, G., Jay, S. C., Nelson, D., & Gates, J. (1995). *A guide to managing bees for crop pollination*. Guelph, Ontario: Canadian Association of Professional Apiculturalists.

SENASICA (Servicio Nacional de Sanidad, Inocuidad y Calidad Agroalimentaria). 2006. Requisitos zoosanitarios para la importación. Retrieved February 25, 2008, from http://148.245.191.4/zooweb/Funcion.aspx.

Southwick, E. E., & Southwick, L. (1992). Estimating the economic value of honey bees (Hymenoptera: Apidae) as agricultural pollinators in the U.S. *Journal of Economic Entomology, 85*, 621–633.

Stephen, W. P. (2003). Solitary bees in North American agriculture: A perspective. In K. Strickler & J. H. Cane (Eds.), *For nonnative crops, whence pollinators of the future?* (41–66). Lanham, MD: Entomological Society of America.

Stout, J. C., Allen, J. A., & Goulson, D. (2000). Nectar robbing, forager efficiency and seed set: Bumblebees foraging on the self-incompatible plant *Linaria vulgaris* Mill. (Scrophulariaceae). *Acta Oecologica, 21*, 277–283.

Sumner, D. A., & Boriss, H. (2006). Bee-conomics and the leap in pollination fees. *Agriculture and Resource Economics Update, 9*, 9–11.

Thomson, D. (2004). Competitive interactions between the invasive European honey bee and native bumble bees. *Ecology, 85*, 458–470.

Thorp, R. W. (2003). Bumble bees (Hymenoptera: Apidae): Commercial use and environmental concerns. In K. Strickler & J. H. Cane (Eds.), *For nonnative crops, whence pollinators of the future?* (21–40). Lanham, MD: Entomological Society of America.

Velthuis, H. H. W., & van Doorn, A. (2006). A century of advances in bumble bee domestication and the economic and environmental aspects of its commercialization for pollination. *Apidologie, 37*, 421–451.

Victorian Scientific Advisory Committee. (2000). Final recommendation on a nomination for listing: The introduction and spread of the large earth bumblebee *Bombus terrestris* L. into Victorian terrestrial environments. Melbourne, Australia: Department of Sustainability and Environment, State of Victoria.

Williams, P. H. (2005). Bombus: *Bumblebees of the world*. Retrieved May 10, 2006, from http://www.nhm.ac.uk/research-curation/projects/bombus/pr.html#ephippiatus.

Wilson, P., & Thomson, D. (1991). Heterogeneity among floral visitors leads to discordance between removal and deposition of pollen. *Ecology, 72*, 1503–1507.

Woodward, D. R. (1996). Monitoring for impact of the introduced leafcutting bee, *Megachile rotundata* (F) (Hymenoptera: Megachilidae), near release sites in South Australia. *Australian Journal of Entomology, 35*, 187–191.

Zimmerman, M., & Cook, S. (1985). Pollinator foraging, experimental nectar-robbing and plant fitness in *Impatiens capensis*. *American Midland Naturalist, 113*, 84–91.

10   Invasive Exotic Plant–Bee Interactions

Karen Goodell

Introduction

Invasions of exotic plants threaten natural ecosystems, exhibiting diverse ecological impacts that are hard to predict (Schmitz et al., 1997; Gordon, 1998; Parker et al., 1999; Ehrenfeld, 2003). They also are economically costly to control and incur millions of dollars in agricultural losses annually (Pimentel et al., 2005). Not surprisingly, most research has focused on the demography and regulation of invasive plant populations (Parker, 2000; Koop & Horvitz, 2005; Hyatt & Araki, 2006). Interactions between invasive plants and their pollinating mutualists can influence the dynamics of both invasive plants and their pollinators (figure 10.1). These interactions are likely important factors in the outcome of plant invasions because more than 90% of angiosperms are animal pollinated (Daily et al., 1997).

Pollinating insects, such as bees, can promote seed production in some invasive plants, which could in turn promote the spread of the invader. Conversely, lack of suitable pollinators could limit the establishment and spread of some non-native plants. Invasions of non-native plants also may affect bee populations and communities directly by contributing floral resources to a community or indirectly by reducing the abundance or diversity of the floral resources available. If pollinators thrive on non-native plants and facilitate seed production, this positive feedback could promote population growth and spread of the invading species. Finally, effects of non-native plants on bee abundance or behavior may indirectly affect the reproduction of native plants that share pollinators with the non-native species through competition or facilitation of flower visitors. Improper pollen transfer by pollinators switching between native and non-native flowers also could reduce the reproductive success of native plants in the presence of the non-native. Invasions of non-native plants could trigger negative feedback on the reproduction of native plants if reductions in native plant densities or diversity also cause declines in the pollinator populations that service native plants (figure 10.1).

Clearly, the particular ecologies of the plant and pollinator species involved influence the outcome of their interactions with invasive plants. Variation in ecological factors, such as mating systems and flowering time of the plants or the diet breadth and reproductive phenology of the pollinators, contribute to a variety of possible responses of native plants and pollinators to plant invasions. Whether the interactions between pollinators and invasive plants are considered positive or negative overall is a value judgment that will depend on the conservation and economic importance of the species involved. Understanding the interactions between pollinators and invasive plants from the pollinator perspective is timely because of concern regarding the status of native pollinator populations (Kearns et al., 1998; Biesmeijer et al., 2006; National Research Council 2006). Plant invasions can be viewed in the context of other research on the factors influencing pollinators (e.g., Steffan-Dewenter & Tscharntke, 1999; Goulson et al., 2002; Kremen et al., 2002, 2004; Thomson, 2004; Cane et al., 2006; McFrederick & LeBuhn, 2006).

Here I look for generalizations that could help predict the outcome of interactions between invasive plants and bees and highlight directions for further research. I start with the role of pollinators in promoting invasions, focusing on data published since this topic was last reviewed (Richardson et al., 2000). I also contribute a new data set on the reproduction of a selection of non-native plants in the United States. I then examine plant invasions from a pollinator perspective. This perspective extends beyond previous analyses of invasive plant-pollinator interactions to consider potential impacts of plant

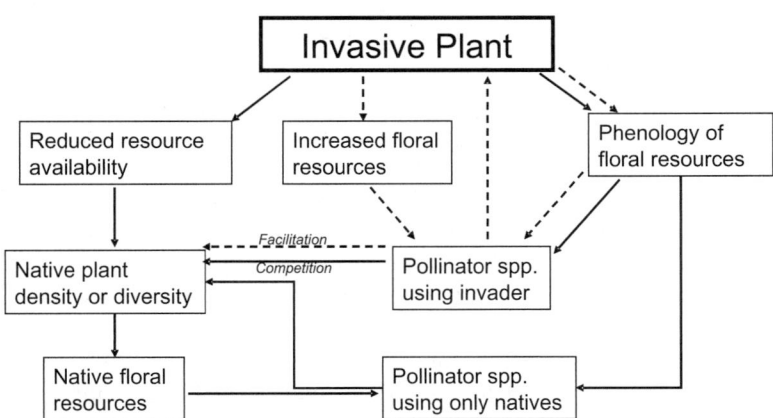

Figure 10.1 Conceptual model showing interactions between invasive plants, native plants, and shared pollinators. Solid lines indicate negative interactions and dashed lines, positive interactions.

invasions on pollinator populations and communities. Finally, I examine indirect effects of invasive plants on native plants to help guide conservation managers, as well as direct future research.

Pollination of Invasive Plants by Pollinators

Many potentially invasive non-native plants are likely to benefit from pollinators because an estimated 80–90% of flowering plants are facultatively or obligately animal pollinated (Brown, 1990; Daily et al., 1997). To what extent does inadequate pollination by resident pollinators constrain the invasions of nonindigenous plants? Conversely, to what extent do pollinators facilitate plant invasions by promoting seed set? The answers to these questions depend on the reproductive biology of the invasive or potentially invasive exotic plants, as well as on their demography. Where strong density dependence exists and population growth is not limited by seed set, increasing seed production will not influence the spread of the invader.

Baker's Law predicts that successful colonizing plants should reproduce vegetatively or apomictically (asexually produced seed) or via self-fertilization, especially if fertilization can be achieved without the aid of biotic pollinators, because these traits provide reproductive assurance to individuals in small founding populations (Baker, 1955, 1965). The local spread of such species would then depend little on pollinators. If animal-mediated pollination is required to set seed by otherwise vegetatively reproducing species, then pollinators might facilitate spread to new patches. Therefore, we expect to see higher proportions of these traits among invasive plants than among plants at large. Many invasive species can produce seed apomictically or autogamously (within-flower self-fertilization) or spread vegetatively, but quantitative data regarding the expression of various reproductive modes of invasive plants are rare, and researchers disagree regarding the importance of animal-mediated pollination (Richardson et al., 2000). One characterization of weedy plants in the United States found that a majority rely on sexual reproduction, with no differences in the frequency of vegetative reproduction between native and exotic weeds (Sutherland, 2004). Invasive plants that require pollinators to set seed are underrepresented in large data sets, which suggests that this trait is disadvantageous in the invasion process (Daehler, 1998). In an explicit test of Baker's Law, Rambuda and Johnson (2004) found high levels of autogamy among exotic weeds in South Africa, with all species showing at least some ability to set seed in the absence of pollinators. None of the species studied exhibited self-incompatibility. Van Kluenen and Johnson's (2007) results also support Baker's Law but suggest that lack of compatible mates, rather than pollinators, limit the distribution of invasive plants in the United States. Invasive plants capable of autonomous pollination had larger ranges than those requiring pollinators, but among species requiring pollinators to reproduce, self-incompatible species occurred in fewer states than self self-compatible species. Furthermore, self-compatible species capable of autonomous pollination had attained similar ranges to those without this ability, suggesting adequate pollinator service. Their data set included non-native species to the United States of European origin only, and distribution was quantified coarsely as presence/absence at the level of states. It does not necessarily reflect patterns of establishment or other measures of invasion success.

To further explore the likely importance of pollinators in determining the outcome of plant invasions into natural areas in the United States, I compiled data on the reproductive modes, mating systems, and pollinator interactions of a subset of current exotic weed species. I randomly sampled 60 species from a list of 1,037 invasive weeds in natural areas generated by the U.S. National Park Service (Swearington, 2006). For each of these 60 species, I searched the current literature using the ISI Science Citation Index Expanded (1980–2007), and two web-based search engines, www.scholar.google.com and www.google.com, for information on the reproduction or pollination of the species. In some cases, I inferred pollinator interactions from personal observations or closely related species. I found published, reliable unpublished, or inferred data on mode of pollination on more than 93% of the plants on my list. The majority of plants exhibited insect pollination; wind pollination was less than half as common (13.3% of wind-pollinated plants were grasses; figure 10.2). Less information was available for the mating systems of invasive angiosperms. Data on mating systems of insect-pollinated species were unavailable for nearly 70%. Self-incompatibility among invasive weeds was slightly less common than self-compatibility (figure 10.3). One species demonstrated autogamy (the ability to self-fertilize in the absence of pollinators), but reliable information

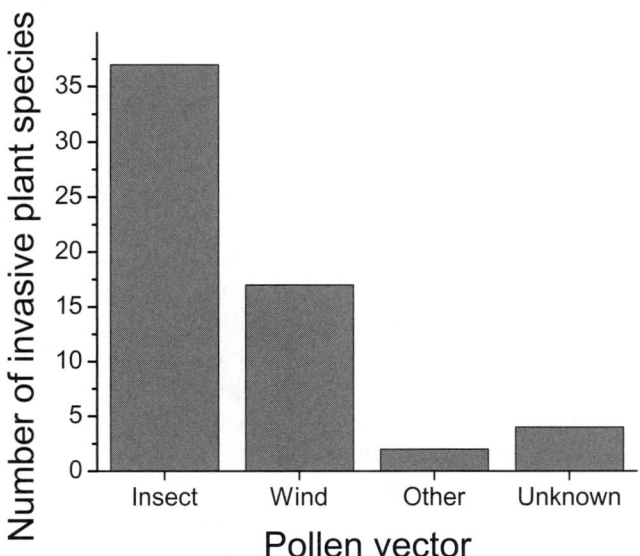

Figure 10.2 The number of invasive plant species that are pollinated by different vectors. The "other" category includes birds and water. Data are a random sample ($n = 60$) taken from WeedsUS, a database of weeds of natural areas in the United States. See text for methods used to generate these data.

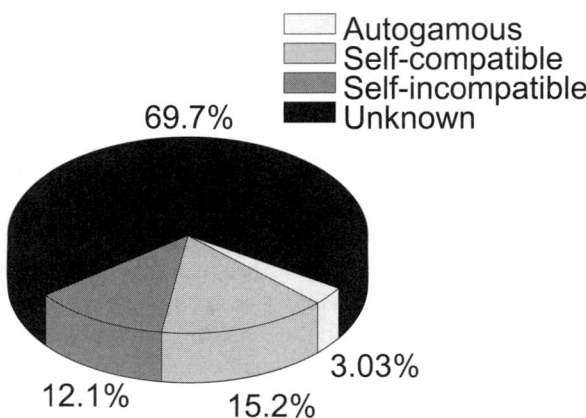

Figure 10.3 Percentage of insect-pollinated invasive plants ($n = 37$; see figure 10.2) that exhibit three different mating strategies. Autogamous plants are self-compatible and can self-fertilize in the absence of pollinators. See text for explanation of database.

was unavailable on how many species had been tested. The number of self-compatible species that require animal pollinators to set seed, then, is unknown. I found no information on apomixis among the 37 insect-pollinated species.

If the plants for which I obtained data indeed represent the pool of non-native invasive plants, it appears that the majority of non-native plant species in natural areas of the United States derive some reproductive benefit from their interactions with pollinators. Bees were the most commonly cited visitor species; 96% of the insect-pollinated species were reported attractive to bees. Pollinator visitation is especially important in the case of plants that cannot set seed without pollinators. Nearly half of the insect-pollinated species for which I found data exhibit self-incompatibility, and so are likely to require pollinators to set seed. Furthermore, only 3% of the self-compatible species demonstrated the ability to set seed without the assistance of pollinators, suggesting that many self-compatible weed species benefit from pollinator visitation. To quantify the role that pollinators play in the reproduction of invasive plants in the United States would require knowledge of the reproductive systems of most invasive plants. An understanding of the basic reproductive biology of invasive plants may also provide an insightful key to their control (e.g., Elam et al., 2007). Collection of reproductive and mating system data should be a top priority among plant ecologists.

Most Invaders Are Preadapted to Resident Pollinators

Non-native plants that are obligately outcrossing must be pollinated by resident pollinator species in the new region if they are to succeed. Most plants can be pollinated effectively by a variety of pollinators (Waser et al., 1996). Those non-native invasive plants requiring pollination are visited by resident generalist pollinators in the invaded range (Richardson et al., 2000; Memmott & Waser, 2002; Olesen et al., 2002; Goulson, 2003; Goulson & Hanley, 2004; Morales & Aizen, 2006). Examples of invaders with highly specific pollinator requirements are rare (but see Nadel et al., 1991; Richardson et al., 2000; and references therein).

Nevertheless, sometimes the presence of non-native pollinator species appears important to the success of invasions. The honey bee, *Apis mellifera*, has been implicated in assisting the invasion of non-native plant species in areas in which honey bees are also non-native. Honey bees' high abundances and broad diets combined with their broad geographic distribution in areas outside their native range likely contribute to the frequency of their interactions with exotic plant species. For example, the Australian invasive plant *Lantana camara* relies on honey bees for pollination because its corolla tube is too long for the primarily short-tongued native species (Goulson & Derwent, 2004). Documented positive relationships between honey bee abundance and fruit set of the invader suggest that honey bees may facilitate the population growth of the invader (Goulson & Derwent, 2004). Barthell et al. (2001) posited that honey bees have facilitated the invasion of non-native yellow star thistle, *Centaurea solstitialis*, in the state of California. Not only were honey bees the most frequent visitors to flowers, but their exclusion from flowers also significantly reduced seed production. Whether honey bees were responsible for the invasion of yellow star thistle depends on whether native visitors might pollinate this species adequately in regions in which honey bees are not the numerically dominant pollinator. Interestingly, the non-native bee species *Megachile apicalis* that specializes on yellow star thistle did not appear to contribute substantially to its seed set, despite its occurrence at some of the study sites (Barthell et al., 2001). Because most invasive plants receive pollinator services from multiple species, their interactions with any one bee species are likely to be weak (Traveset & Richardson, 2006). It is unlikely, then, that non-native bees promote plant invasions in general.

One important exception may be cases in which humans introduce non-native pollinators into an area with a depauperate pollinator fauna. In these situations, the introduced pollinator may represent a new functional group and could pave the way for plant invasions that would otherwise be limited by pollinator availability. The pollination of *Lantana* by introduced honey bees in Australia, discussed earlier, is one example. Another example is the introduction of bumble bees (*Bombus terrestris*) in Australia, New Zealand, and Tasmania, areas that have few native long-tongued bee species (Hanley & Goulson, 2003). In these regions, bumble bees appear to promote the invasion of weed species from Europe and North America that are adapted to long-tongued bee pollinators. In Tasmania, where bumble bees are just becoming established, seed production in incipient invasions of non-native *Lupinus arboreus* was positively correlated with bumble bee visitation rates (Stout et al., 2002). In New Zealand, where non-native *Bombus terrestris* has become widespread, *L. arboreus* is already a problematic invader (Stout et al., 2002). *Cytisus scoparius*, a native of Europe, requires pollinators strong enough to trip its flowers and is pollinated primarily by bumble bees in its native range (Simpson et al., 2005). Because *C. scoparius* spreads only by seed, increased densities of non-native bumble bees could augment its reproductive success and promote its spread in Australia, where it has been introduced. Bumble bee pollination may especially promote spread of *C. scoparius* in cooler areas where researchers have hypothesized that the greater forces needed to open flowers limit the effectiveness of honey bees as pollinators (Parker & Haubensak, 2002). Interestingly, all of these examples of invasive plants are of the Fabaceae: Papilionoideae. The observed pattern of synergism between introduced bumble bees and the papilionoid legumes in areas such as Australia and New Zealand provides predictive power that could be used to develop policy for new introductions and strategies for management. For example, introduction of non-native bumble bees or honey bees could promote spread of the exotic plant gorse, *Ulex europeaus*, in areas such as the Channel Islands of New Zealand, where honey bees and bumble bees have not yet established (MacFarlane et al., 1992).

Pollen Limitation in Invasive Species

Does bee visitation limit seed set of invasive plant species? Pollen-limited seed set has been reported for the North American invaders *C. scoparius* and *Genista monspessulana* (Parker, 1997; Parker & Haubensak, 2002). Pollen limitation in *C. scoparius* also varied with other factors such as resources and florivory (Parker & Haubensak, 2002). The invasive vine kudzu, *Puraria montana*, also exhibited variation among populations in pollen limitation of seed set (Forseth & Innis, 2004). *Lonicera japonica*, an invader in North America that exhibits adaptations for hawkmoth pollination, has severely pollen-limited seed set (Larson et al., 2002). In *L. japonica*, low visitation by nocturnal hawkmoths and diurnal bees, as well as their inefficiency, contributed to low seed set (Larson et al., 2002). In contrast, Rambuda and Johnson (2004) found no evidence for pollen limitation of seed set in 17 South African invasive weeds, but most were capable of autogamy, and all were self-compatible and, therefore, are not likely to exhibit pollen limitation of seed set even when pollinator visitation to flowers is infrequent.

Pollen limitation of seed production only can limit the spread of invasive plants if seed production is linked to the demography of the plant. For few invaders do we know the extent to which seed set limits population growth. Parker (1997) found that *C. scoparius* population growth was not predicted by fecundity. Given the number of factors that could decouple seed set from population growth, it seems likely that further research will show variation among localities in the importance of seed set to population growth. Parker and Haubensak (2002) point out, however, that spread of nonclonally reproducing invasive plants will always be seed and dispersal limited at the margins of the invasions and at the start of new local invasions; and, therefore, lack of pollinators may control aspects of many invasions.

Poor seed set due to pollen limitation may have the greatest impact on the population growth of small founding populations (Taylor & Hastings, 2005) that cannot attract adequate pollinators. No studies have examined this problem for an incipient invasion but have instead focused on already successful invaders in relatively large populations. Our understanding of the role that pollinators play in limiting invasions, then, would benefit from knowledge of the mating systems and pollinator associations of all introduced species, including those that become successful, as well as those that fail to establish. Failed introductions of exotic species are common, being estimated as high as 90% (Williamson & Fitter, 1996). For plants, however, data on rates of failed introductions are rare, and we have almost no information about the mechanisms behind those failures. In the absence of such data, pollinator attributes for invasive non-native plants could be compared with those of non-invasive, non-native plants. Furthermore, experimental studies of the colonization process by non-native plants that measure the role of pollen limitation of seed set in the population growth rates will help assess the contribution of bees early in the invasion process.

Invasive Plant Effects on Bees

The contribution of pollinating mutualists is increasingly recognized as an important factor in the invasion process for some non-native plants, yet the direct impacts of invasive plants on populations and communities of pollinators (figure 10.1) remain largely unstudied. From a pollinator perspective, many invasive plants may increase the carrying capacity of the habitat by providing an unprecedented resource boon. My survey of the WeedsUS database (see preceding discussion) indicated that 96% of the insect-pollinated invasive species and another 12% of the wind-pollinated species offer pollen, nectar, or both resources to bees. Because of the close link expected between the density of bees and the availability of floral resources (e.g., Potts et al., 2003), non-native invasive plants could potentially change the density and composition of pollinators.

For example, Schurkens and Chittka (2001) report that the abundant flowers of the invasive crucifer *Bunias orientalis* could be important in supporting pollinator populations, despite paltry nectar offerings on a per flower basis. Rich nectar rewards per flower, on the other hand, explain why non-native *Impatiens glandulifera* lures bees away from native flowers; it had a higher rate of nectar secretion than any native plant on record in the invaded region (Chittka & Schurkens, 2001). Both of these invasive plant species potentially exert a large effect on pollinator populations via augmenting resources, but without knowing how the abundance and distribution of floral resources have changed following the invasion, it is difficult to predict how pollinator communities will respond.

Not only may the quantity of nectar and pollen differ postinvasion, but the quality (e.g., nectar sugar concentration or pollen protein content) may differ, as well. Furthermore, flowering phenology and flower morphology of the invader relative to the native plants will influence resource availability to pollinators (figure 10.1). Shifts in resource availability and composition occur as a result of the presence of the new plant species and the absence of native species that have been displaced by the invader. Such changes in the quality and availability of floral resources following invasions could cause changes in pollinator community composition. An example of a plant that shifts both the abundance and the temporal distribution of floral resources is the North American invader *Lythrum salicaria*, which has invaded calcareous fens in New Jersey. It flowers prolifically in July and August, offering more floral resources than fen communities that have not been invaded. Invaded fens had fewer floral resources in early spring than uninvaded fens, however, because *L. salicaria* displaces native plants that bloom in the spring (Goodell, unpublished data). Bee samples that were passively collected in the early spring revealed that invaded sites had comparable bee abundances but 50% lower richness than uninvaded sites (Goodell, 2003). During the *L. salicaria* bloom, pollinator densities on flowers did not vary with invasion status, but composition did. Large-bodied bees (*Apis mellifera* and *Bombus* spp.) were more abundant in invaded sites, whereas small-bodied native bees (>50 different species) were more abundant in uninvaded sites (Goodell, 2003). In this system, changes in floral communities had a larger effect on the composition of pollinator communities than on pollinator abundance.

Non-native plants have been hypothesized to promote the invasion of non-native bees. The "invasional meltdown" hypothesis predicts that positive interactions between two or more invasive species, such as a plant and a pollinator, could accelerate the replacement of the native species with exotic species (Simberloff & Von Holle, 1999). Several researchers have asked whether densities of non-native plants were positively associated with densities of non-native bees, but their data are too scarce to generalize on the likelihood of plant-pollinator-driven "meltdowns" of bee communities. The positive correlations between the abundance of introduced crops, such as *Trifolium pratense* and *Lotus corniculatus*, and two introduced bumble bee species reported in New Zealand appear to support the idea that exotic plants can bolster exotic bee populations (Hanley & Goulson, 2003). On the other hand, Morales and Aizen (2002) found no consistent relationship between non-native honey bee or bumble bee density and non-native flower density in South American temperate dry forests. These studies fall short of demonstrating plant-bee-driven invasional meltdowns. Studies demonstrating that positive interactions between non-native plants and non-native bees negatively affect the population dynamics of native species are still needed (Simberloff, 2006).

Research into the spread of specialist pollinator species with their introduced host plants might offer a clearer picture of how positive interactions between non-native pollinators and invasive plants influence non-native bee distributions. For example, exotic wasps that are specialist pollinators of *Ficus* spp. have established and spread into parts of the United States following the establishment of their host plant. This establishment documents a positive feedback between plant-pollinator invaders (Nadel et al. 1991; Corlett, 2006). The spread of the squash bees in the genus *Peponapis* and *Xenoglossa* through the Americas as human civilizations extended the range of their cultivated host plants (*Cucurbita* spp.) is another example (Hurd et al., 1974). The invasion of yellow star thistle into the western United States has allowed proliferation and spread of the *Centaurea* leaf-cutting bee, *Megachile apicalis*, an apparent oligolege on *Centaurea* spp. (Barthell et al., 2003; Stephen, 2003). In addition, Cane (2003) provides a list of non-native bees in North America that includes at least 10 species that exhibit some level of host specialization or preference for non-native plants. These species would be a logical point of departure for investigations on the effects of non-native host plant spread and abundance on the spread of non-native bees.

Changes in floral resource abundance and availability are not the only mechanisms for impact of invasive species on native bees. Invasions could also alter the availability of nesting habitat. The availability and distribution of bare ground, certain types of ground cover, hollow stems of shrubs, and dead logs may all change as a consequence of floral turnover associated with the invasion of non-native plant species. Nesting habitat is especially important in bees because they are central-place foragers and require both food and nest sites within their foraging range. Finally, bee pollinators could be affected by physical or structural changes to a habitat, such as those that affect light availability or temperature conditions, which in turn could influence bee flight and foraging (Herrera, 1995, 1997; Hansen & Totland 2006).

In summary, the hypothesis that invasive plants boost, diminish, or alter the composition of pollinator communities has limited evidence and is inadequately addressed by research. Given the value of bees and other pollinators in sustaining native plants, the effect of plant invasions on bee populations and communities is worth investigating further. The impacts of invasive plants on bees could be considered a special case of a more general issue: the effects of changing plant communities on bees. Other human-driven alterations to plant communities and floral resources could also affect bee communities, such as urbanization (Goulson et al., 2002; Cane et al., 2006), agricultural activity (Kremen et al., 2002; Westphal et al., 2003), and fire (Potts et al., 2003). What are the consequences of these changes in bee communities? We know that spatial and temporal variation in flower-visitor composition abounds in natural populations of plants (e.g., Herrera, 1988; Moeller, 2005). An important question is whether the effects of invasive plants on pollinator communities are important to the reproduction of native plants despite all of the underlying variation in pollinator services expected.

Pollinator-Mediated Effects of Invasive Plants on Native Plant Reproduction

The negative effects of invasive plants on native plants through competition for resources such as water or nitrogen affect those native plants growing in a localized area around the invasion. When the resource in question is the services of shared pollinators, the invader's sphere of influence can extend to more distant plants and even to plants in other habitats. In theory, pollinator-mediated interactions with non-native invasive plants could disrupt or enhance the pollination of native plant species (Traveset & Richardson, 2006). Because most plants and most pollinators, including most bee species, are generalists (Waser et al., 1996), most invasive plants will interact with many resident pollinator species. Consequently, native plants should frequently share pollinators with invasive plants, creating novel, indirect interactions between native and invasive plants. In fact, this pattern has been observed in nature (Memmott & Waser, 2002; Olesen et al., 2002; Morales & Aizen, 2006). These novel pollinator-mediated indirect interactions between native and invasive plants could reduce pollinator service to native plants under some conditions, but the opposite is also possible (Ashman et al., 2004; Knight et al., 2005).

Competition for pollinators between invaders and native plant species is expected when the invader offers equivalent or richer floral rewards and likely reflects a change in pollinator behavior. Facilitation of pollination by invasive plants is expected when the invader offers rewards complementary to those of native species that help sustain pollinators during periods of resource scarcity. The strength of pollinator-mediated effects of invaders on native plant reproduction should increase with overlap in pollinator species. Most studies that have examined pollinator-mediated interactions between invasive and native plants have done so for one or a few native plant species. Generalizations about how often pollinator-mediated impacts affect native plants and what sorts of native plants are most vulnerable would be helpful to assess the magnitude of the problem relative to other impacts of invasive plants.

Nine studies (one unpublished) assess pollinator-mediated impacts of eight different invasive plant species on reproduction of 19 native plant species. Visitation rates by bees differed significantly for native plants in the presence and absence of the invasive plant in 9 of 14 cases. Seven native plants received fewer bee visits in the presence of the invasive plant than in its absence, indicating competition for pollinators (Chittka & Schurkens, 2001; Aigner, 2004; Ghazoul, 2004; Moragues & Traveset, 2005; Larson et al., 2006). Only two native plant species received more visits in the presence of the invader plant, indicative of facilitation, and in both cases this effect was inconsistent among years (Moragues & Traveset, 2005). The presence of invasive plants had no effect on the visitation rates to native plants of five species (Grabas & Laverty, 1999; Brown et al., 2002; Moragues & Traveset, 2005; Goodell, unpublished data).

Although competition for bee visitation was the most common type of interaction observed, it did not always reduce plant reproduction (Ghazoul, 2002, 2004). The quality of pollen deposited by flower visitors, as measured by the number of conspecific pollen grains transferred to stigmas, forms the functional link between visitation rate and seed set. If native plants receive sufficient pollen to fertilize ovules in the presence of the invader, reproduction may not be affected, even if pollen of the invader is also transferred (Brown et al., 2002; Ghazoul, 2004; Moragues & Traveset, 2005; also see chapter 11, this volume). Only 2 native species of 11 tested were reported to have reduced pollen quality (proportion of conspecific pollen deposited) in the presence of the invader. One study showed negative effects of invader pollen on native plant reproduction (Brown & Mitchell, 2001), and the other did not (Moragues & Traveset, 2005).

Despite the frequency of competition for pollinators with highly rewarding invaders, relatively little data indicate that competition for visitors reduces seed set in natives. In addition, competitive effects have not been found consistently across years (Moragues & Traveset, 2005; Larson et al. 2006). The conclusion that pollinator-mediated effects on native plants are not an important aspect of plant invasions, however, is premature. Relatively few pairs of invasive and native plants have been investigated, and some of these studies did not examine seed set. Even if competition is rare, careful attention to possible negative effects on native plants of conservation importance is warranted. To that end, developing some rules of thumb to identify those native plants most likely to suffer negative effects would be useful for conservation managers and future investigators. Here I examine evidence for the predictions that: (1) the likelihood of strong pollinator-mediated interactions should increase as overlap in pollinator species increases and (2) invasive plant-native plant species pairs with similar floral morphology should compete for pollinators more intensely than those with distinct morphologies because they are expected to have similar suites of pollinator species.

Using the preceding studies as examples, the average overlap in pollinators among invaders and natives in cases with significant pollinator-mediated interactions was 69.3% ($SD = 23.7\%$, $n = 4$), significantly greater than the 35.1% ($SD = 23.6\%$, $n = 5$) for those lacking significant pollinator-mediated interactions (two-sample t-test with arcsine square-root transformed data, $t = -2.03$, p one-tailed $= 0.04$). The limitations of these data are that the studies did not all assess overlap to the same taxonomic level, with some identifying pollinators to species and others grouping them by genus, higher taxonomic levels, or even taxonomically independent categories. Nevertheless, these studies tentatively support the general hypothesis that greater overlap increases likelihood of finding significant pollinator-mediated effects.

Relatively few studies have measured pollinator-mediated interactions between nonindigenous plants and native plants of similar versus dissimilar morphology. Grabas and Laverty (1999) found no consistent evidence of competition or facilitation for pollinators between invasive *L. salicaria* and native flowers with relatively generalized versus specialized pollinator interactions. *Lythrum salicaria*, however, attracts a very wide variety of insect taxa. Theoretically, invasive species with more specialized floral morphology and pollinator associations may compete more intensely with native plants possessing similar floral morphology. Available empirical evidence does not support this hypothesis, however. I scored pairs of native and non-native plants as similar or different based on whether they exhibited zygomorphic or actinomorphic flowers. Data on visitation and seed set show that 4 of 16 native species had similar floral morphology to the invader, yet none of these exhibited competition for visits or reduced seed set (Moragues & Traveset, 2005; Larson et al., 2006). Similar patterns are found for seed set data (Moragues & Traveset, 2005; Larson et al., 2006). I also found no evidence that morphological similarity influenced the likelihood of facilitative interactions. Similarity in floral morphology, then, does not appear to strongly influence the probability that invasive plants will compete for or facilitate pollination of natives, at least at the coarse level at which floral similarity was considered here. Possibly, extremely similar floral structure increases the chances of competition. One study that examined competition between congeners found substantial competition that reduced visitation and seed set in the native species (Brown & Mitchell, 2001).

The rule of thumb that emerges from these results is that pollinator-mediated impacts of an invasive plant on native plants should focus on native-invasive pairs with a broad pollinator overlap (>65%). Second, similarity of floral morphology between potentially competing plants does not appear important to competition. Further research will likely reveal other rules of thumb.

Conclusions

Although many invasive plants are visited by bees and may even require bees to pollinate them, the degree to which bees directly facilitate plant invasions is not well documented. The stage at which bees seem most likely to exert influence is the incipient invasion stage for plants that require pollinators to set seed and do not reproduce vegetatively. More studies of reproduction and pollination during these early stages would advance our understanding of bee impacts on the invasion process.

Mature invasions represented by a numerically dominant invader that offers attractive, rewarding flowers seem likely to influence bee populations, but the nature of the interaction will depend on how the invader compares with the preinvasion floral community. Simple comparisons of floral reward structure before and after invasion or in the presence and absence of the invader would help to develop predictions about their effects on bee communities. My studies of *L. salicaria* invasions document broad changes in the bee community composition associated with invasion. These changes appear to result from different bee species responding to different parts of the floral community. Given the ecological diversity of bees, this result is not surprising. No studies have examined how alterations in nesting habitat or other habitat features following invasions may influence bee communities. Further research is needed in these areas.

Pollinator-mediated indirect effects of invasive plants manifested as competition or facilitation for pollinators have been hypothesized repeatedly, but little empirical evidence documents their importance to native plants. Rather than more studies on a single invader-native pair, researchers should focus on testing general predictions regarding the type of native plants likely to suffer competition, the spatial scale over which these interactions take place, or contexts in which an invasion is likely to significantly alter bee behavior in ways that could affect native plant reproduction. My synthesis of the current studies indicates that pollinator overlap, but not floral morphology, influence the strength of these indirect interactions.

Gaining a better understanding of the effects of invasive plants on pollinator communities and their potential indirect effects on native plants not only helps us assess the conservation implications of invasion but also is important when considering removals of invasive plants. Control measures that rapidly eliminate an important food source for bees could have unintended negative repercussions for bee populations and the native plants they serve. Restoration work that considers establishment of alternative floral resources in the short and long term may avoid pitfalls that could destroy an important part of the plant-bee food web. Dealing with the invasive plant crisis could benefit from abandoning the view of invasive plants as noxious pests that have only negative effects and adopting an ecological view that considers both the positive and negative interactions with invasive plants and how they influence the target community.

Acknowledgments

The research on *Lythrum salicaria* was funded by a D. H. Smith Conservation Research Postdoctoral Fellowship. This chapter benefited from comments on earlier drafts from Ingrid Parker, Amy McKinney, Kristin Mercer, Rosalind James, and Theresa Pitts-Singer.

References

Aigner, P. A. (2004). Ecological and genetic effects on demographic processes, pollination, clonality and seed production in *Dithyrea maritime*. *Biological Conservation, 116*, 27–34.

Ashman, T. L., Knight, T. M., Steets, J. A., Amarasekare, P., Burd, M., Campbell, D. R., et al. (2004). Pollen limitation of plant reproduction: Ecological and evolutionary causes and consequences. *Ecology, 85*, 2408–2421.

Baker, H. G. (1955). Self-compatibility and establishment after "long-distance" dispersal. *Evolution, 9*, 347–349.

———. (1965). Support for "Baker's law" as a rule. *Evolution, 21*, 853–856.

Barthell, J., Thorp, R., Frankie, G. W., Kim, J. Y., & Hranitz, J. M. (2003). Impacts of introduced solitary bees on natural and agricultural systems: The case of the leafcutting bee *Megachile apicalis* (Hymenoptera: Megachilidae). In K. Strickler & J. Cane (Eds.), *For nonnative crops, whence pollinators of the future?* (151–163). Lanham, MD: Entomological Society of America.

Barthell, J. F., Randall, J. M., Thorp, R. W., & Wenner, A. M. (2001). Promotion of seed set in yellow star-thistle by honey bees: Evidence of an invasive mutualism. *Ecological Applications, 11*, 1870–1883.

Biesmeijer, J. C., Roberts, S. P. M., Reemer, M., Ohlemuller, R., Edwards, M., Peeters, T., et al. (2006). Parallel declines in pollinators and insect-pollinated plants in Britain and the Netherlands. *Science, 313*, 351–354.

Brown, A. D. H. (1990). Genetic characterization of plant mating systems. In A. D. H. Brown, M. T. Clegg, & A. L. Kahler (Eds.), *Plant population genetic resources, breeding, and genetic resources* (145–162). Sunderland, MA: Sinauer.

Brown, B. J., & Mitchell, R. J. (2001). Competition for pollination: Effects of pollen of an invasive plant on seed set of a native congener. *Oecologia, 129*, 43–49.

Brown, B. J., Mitchell, R. J., & Graham, S. A. (2002). Competition for pollination between an invasive species (purple loosestrife) and a native congener. *Ecology, 83*, 2328–2336.

Cane, J. H. (2003). Exotic non-social bees (Hymenoptera: Apoidea) in North America: Ecological implications. In K. Strickler & J. Cane (Eds.), *For nonnative crops, whence pollinators of the future?* (113–126). Lanham, MD: Entomological Society of America.

Cane, J. H., Minckley, R. L., Kervin, L. J., Roulston, T. H., & Williams, N. M. (2006). Complex responses within a desert bee guild (Hymenoptera: Apiformes) to urban habitat fragmentation. *Ecological Applications, 16*, 632–644.

Chittka, L., & Schurkens, S. (2001). Successful invasion of a floral market: An exotic Asian plant has moved in on Europe's river-banks by bribing pollinators. *Nature, 411*, 653–653.

Corlett, R. T. (2006). Figs (*Ficus*, Moraceae) in urban Hong Kong, south China. *Biotropica, 38*, 116–121.

Daehler, C. C. (1998). The taxonomic distribution of invasive angiosperm plants: Ecological insights and comparison to agricultural weeds. *Biological Conservation, 84*, 167–180.

Daily, G. C., Alexander, S., Ehrlich, P. R., Goulder, L., Lubchenco, J., Matson, P. A., et al. (1997). Ecosystem services: Benefits supplied to human societies by natural ecosystems. *Issues in Ecology, 2*, 1–16.

Ehrenfeld, J. G. (2003). Effects of exotic plant invasions on soil nutrient cycling processes. *Ecosystems, 6*, 503–523.

Elam, D. R., Ridley, C. E., Goodell, K., & Ellstrand, N. C. (2007). Population size and relatedness affect the fitness of a self-incompatible invasive plant. *Proceedings of the National Academy of Sciences of the USA, 104*, 549–555.

Forseth, I. N., & Innis, A. F. (2004). Kudzu (*Pueraria montana*): History, physiology and ecology combine to make a major ecosystem threat. *Critical Reviews in Plant Sciences, 23*, 401–413.

Ghazoul, J. (2002). Flowers at the front line of invasion? *Ecological Entomology, 27*, 638–640.

———. (2004). Alien abduction: Disruption of native plant-pollinator interactions by invasive species. *Biotropica, 36*, 156–164.

Goodell, K. (2003). Structure of bee communities in calcareous fens invaded by purple loosestrife compared to uninvaded fens [Abstract]. *Ecological Society of America Annual Meeting*. Savannah, GA: Allen Press.

Gordon, D. R. (1998). Effects of invasive, non-indigenous plant species on ecosystem processes: Lessons from Florida. *Ecological Applications, 8*, 975–989.

Goulson, D. (2003). Effects of introduced bees on native ecosystems. *Annual Review of Ecology and Systematics, 34*, 1–26.

Goulson, D., & Derwent, L. C. (2004). Synergistic interactions between an exotic honey bee and an exotic weed: Pollination of *Lantana camara* in Australia. *Weed Research, 44*, 195–202.

Goulson, D., & Hanley, M. E. (2004). Distribution and forage use of exotic bumble bees in South Island, New Zealand. *New Zealand Journal of Ecology, 28*, 225–232.

Goulson, D., Hughes, W. O. H., Derwent, L. C., & Stout, J. C. (2002). Colony growth of the bumble bee, *Bombus terrestris*, in improved and conventional agricultural and suburban habitats. *Oecologia, 130*, 267–273.

Grabas, G. P., & Laverty, T. M. (1999). The effect of purple loosestrife (*Lythrum salicaria* L., Lythraceae) on the pollination and reproductive success of sympatric co-flowering wetland plants. *Ecoscience, 6*, 230–242.

Hanley, M. E., & Goulson, D. (2003). Introduced weeds pollinated by introduced bees: Cause or effect? *Weed Biology and Management, 3*, 204–212.

Hansen, V. I., & Totland, Ø. (2006). Pollinator visitation, pollen limitation, and selection on flower size through female function in contrasting habitats within a population of *Campanula persicifolia*. *Canadian Journal of Botany-Revue Canadienne De Botanique, 84*, 412–420.

Herrera, C. M. (1995). Microclimate and individual variation in pollinators—flowering plants are more than their flowers. *Ecology, 76*, 1516–1524.

Herrera, C. M. (1997). Thermal Biology and foraging responses of insect pollinators to the forest floor irradiance mosaic. *Oikos, 78*, 601–611.

Herrera, C. M. (1988). Variation in mutualisms: The spatio-temporal mosaic of a pollinator assemblage. *Biological Journal of the Linnean Society, 35*, 95–125.

Hurd, P. D., Linsley, E. G., & Michelbacher, A. E. (1974). Ecology of the squash and gourd bee, *Peponapis pruinosa*, on cultivated cucurbits in California (Hymenoptera: Apoidea). *Smithsonian Contributions to Zoology, 168*, 1–17.

Hyatt, L. A., & Araki, S. (2006). Comparative population dynamics of an invading species in its native and novel ranges. *Biological Invasions, 8*, 261–275.

Kearns, C. A., Inouye, D. W., & Waser, N. M. (1998). Endangered mutualisms: The conservation of plant-pollinator interactions. *Annual Review of Ecology and Systematics, 29,* 83–112.

Knight, T. M., Steets, J. A., Vamosi, J. C., Mazer, S. J., Burd, M., Campbell, D. R., et al. (2005). Pollen limitation of plant reproduction: Pattern and process. *Annual Review of Ecology Evolution and Systematics, 36,* 467–497.

Koop, A. L., & Horvitz, C. C. (2005). Projection matrix analysis of the demography of an invasive, nonnative shrub (*Ardisia elliptica*). *Ecology, 86,* 2661–2672.

Kremen, C., Williams, N. M., Bugg, R. L., Fay, J. P., & Thorp, R. W. (2004). The area requirements of an ecosystem service: Crop pollination by native bee communities in California. *Ecology Letters, 7,* 1109–1119.

Kremen, C., Williams, N. M., & Thorp, R. W. (2002). Crop pollination from native bees at risk from agricultural intensification. *Proceedings of the National Academy of Sciences of the USA, 99,* 16812–16816.

Larson, D. L., Royer, R. A., & Royer, M. R. (2006). Insect visitation and pollen deposition in an invaded prairie plant community. *Biological Conservation, 130,*148–159.

Larson, K. C., Fowler, S. P., & Walker, J. C. (2002). Lack of pollinators limits fruit set in the exotic *Lonicera japonica*. *American Midland Naturalist, 148,* 54–60.

MacFarlane, R. P., Grundell, J. M., & Dugdale, J. S. (1992). Gorse on the Chatam Islands seed formation, arthropod associations and control. In *Proceedings of the 45th New Zealand Plant Protection Conference* (251–255). Hastings: New Zealand Plant Protection Society.

McFrederick, Q. S., & LeBuhn, G. (2006). Are urban parks refuges for bumble bees *Bombus* spp. (Hymenoptera: Apidae)? *Biological Conservation, 129,* 372–382.

Memmott, J., & Waser, N. M. (2002). Integration of alien plants into a native flower-pollinator visitation web. *Proceedings of the Royal Society of London: Series B. Biological Sciences, 269,* 2395–2399.

Moeller, D. A. (2005). Pollinator community structure and sources of spatial variation in plant-pollinator interactions in *Clarkia xantiana* ssp *xantiana*. *Oecologia, 142,* 28–37.

Moragues, E., & Traveset, A. (2005). Effect of *Carpobrotus* spp. on the pollination success of native plant species of the Balearic Islands. *Biological Conservation, 122,* 611–619.

Morales, C. L., & Aizen, M. A. (2002). Does invasion of exotic plants promote invasion of exotic flower visitors? A case study from the temperate forests of the southern Andes. *Biological Invasions, 4,* 87–100.

———. (2006). Invasive mutualisms and the structure of plant-pollinator interactions in the temperate forests of north-west Patagonia, Argentina. *Journal of Ecology, 94,* 171–180.

Nadel, H., Frank, J. H., & Knight, R. J. (1991). Escapees and accomplices: The naturalization of exotic (*Ficus*) and their associated faunas in Florida. *Florida Entomologist, 75,* 29–38.

National Research Council. (2006). *Status of pollinators in North America*. Washington, DC: National Academies Press.

Olesen, J. M., Eskildsen, L. I., & Venkatasamy, S. (2002). Invasion of pollination networks on oceanic islands: Importance of invader complexes and endemic super generalists. *Diversity and Distributions, 8,* 181–192.

Parker, I. M. (1997). Pollinator limitation of *Cytisus scoparius* (Scotch broom), an invasive exotic shrub. *Ecology, 78,* 1457–1470.

———. (2000). Invasion dynamics of *Cytisus scoparius*: A matrix model approach. *Ecological Applications, 10*, 726–743.

Parker, I. M., & Haubensak, K. A. (2002). Comparative pollinator limitation of two non-native shrubs: Do mutualisms influence invasions? *Oecologia, 130*, 250–258.

Parker, I. M., Simberloff, D., Lonsdale, W. M., Goodell, K., Wonham, M., Kareiva, P. M., et al. (1999). Impact: Toward a framework for assessing the ecological effects of invaders. *Biological Invasions, 1*, 3–19.

Pimentel, D., Zuniga, R., & Monison, D. (2005). Update on the environmental and economic costs associated with alien-invasive species in the United States. *Ecological Economics, 52*, 273–288.

Potts, S. G., Vulliamy, B., Dafni, A., Ne'eman, G., O'Toole, C., Roberts, S., et al. (2003). Response of plant-pollinator communities to fire: Changes in diversity, abundance and floral reward structure. *Oikos, 101*, 103–112.

Rambuda, T. D., & Johnson, S. D. (2004). Breeding systems of invasive alien plants in South Africa: Does Baker's rule apply? *Diversity and Distributions, 10*, 409–416.

Richardson, D. M., Allsop, N., D'Antonio, C. M., Milton, S. J., & Rejmánek, M. (2000). Plant invasions: The role of mutualisms. *Biological Reviews, 75*, 65–93.

Schmitz, D. C., Simberloff, D., Hofstetter, R. H., Haller, W., & Sutton, D. (1997). The ecological impact of nonindigenous plants. In D. Simberloff, D. C. Schmitz, & T. C. Brown (Eds.), *Strangers in paradise: Impact and management of nonindigenous species in Florida* (39–61). Washington, DC: Island Press.

Schurkens, S., & Chittka, L. (2001). The significance of the invasive crucifer species *Bunias orientalis* (Brassicaceae) as a nectar source for central European insects. *Entomologia Generalis, 25*, 115–120.

Simberloff, D. (2006). Invasional meltdown 6 years later: Important phenomenon, unfortunate metaphor, or both? *Ecology Letters, 9*, 912–919.

Simberloff, D., & Von Holle, B. (1999). Positive interactions of nonindigenous species: Invasional meltdown? *Biological Invasions, 1*, 21–32.

Simpson, S. R., Gross, C. L., & Silberbauer, L. X. (2005). Broom and honey bees in Australia: An alien liaison. *Plant Biology, 7*, 541–548.

Steffan-Dewenter, I., & Tscharntke, T. (1999). Effects of habitat isolation on pollinator communities and seed set. *Oecologia, 121*, 432–440.

Stephen, W. P. (2003). Solitary bees in North American agriculture: A perspective. In K. Strickler & J. H. Cane (Eds.), *For nonnative crops, whence pollinators of the future?* (41–66). Lanham, MD: Entomological Society of America,.

Stout, J. C., Kells, A. R., & Goulson, D. (2002). Pollination of the invasive exotic shrub *Lupinus arboreus* (Fabaceae) by introduced bees in Tasmania. *Biological Conservation, 106*, 425–434.

Sutherland, S. (2004). What makes a weed a weed: Life history traits of native and exotic plants in the USA. *Oecologia, 141*, 24–39.

Swearington, J. (2006). *WeedsUS: Database of plants invading natural areas in the U.S.* Retrieved October 25, 2004, from the U.S. National Parks Service website,http://www.nps.gov/plants/alien/list/all.htm.

Taylor, C. M., & Hastings, A. (2005). Allee effects in biological invasions. *Ecology Letters, 8*, 895–908.

Thomson, D. (2004). Competitive interactions between the invasive European honey bee and native bumble bees. *Ecology, 85,* 458–470.

Traveset, A., & Richardson, D. M. (2006). Biological invasions as disruptors of plant reproductive mutualisms. *Trends in Ecology and Evolution, 21,* 208–216.

Van Kleunen, M., & Johnson, S. D. (2007). Effects of self-compatability on the distribution range of invasive European plants in North America. *Conservation Biology, 21,* 1537–1544.

Waser, N. M., Chittka, L., Price, M. V., Williams, N. M., & Ollerton, J. (1996). Generalization in pollination systems, and why it matters. *Ecology, 77,* 1043–1060.

Westphal, C., Steffan-Dewenter, I., & Tscharntke, T. (2003). Mass flowering crops enhance pollinator densities at a landscape scale. *Ecology Letters, 6,* 961–965.

Williamson, M. H., & Fitter, A. (1996). The characters of successful invaders. *Biological Conservation, 78,* 163–170.

11 Estimating the Potential for Bee-Mediated Gene Flow in Genetically Modified Crops

James E. Cresswell

Introduction

Genetically modified (GM) crops were first released in 1986 (Barber, 1999) and have been in commercial production for 10 years, with a current global area of approved cultivation in excess of 81 million ha across 17 countries (James, 2004), although about 99% of this area lies in four countries, namely the United States, Argentina, Canada, and China. Most of the currently commercialized GM crops are based on introduced genes that confer traits associated with plant welfare (Barber, 1999), such as insect resistance or herbicide tolerance, and none of these pose a proven safety risk to humans through cross-pollination with conventional agriculture. Nevertheless, there has been considerable scientific response (Colwell et al., 1985; Lutman, 1999; Poppy & Wilkinson, 2005) to the public concern about the environmental effects of GM crops and the possible incorporation of GM genes into the human food chain. The concern has arisen particularly in Europe, where the commercial cultivation of GM crops currently is proscribed (Weekes et al., 2005). The prospects for the eventual deployment of GM crops within European agriculture are constrained by the European Union's regulatory thresholds, which restrict the levels of GM adventitious presence in yields from conventional and organic crops. Consequently, the future coexistence of conventional and GM crops may compromise a farmer's ability to certify produce for sale in the potentially lucrative GM-free market in Europe (Belcher et al., 2005). Of the major GM crops, three (soybean, cotton, and canola) are capable of being cross-pollinated by bees (Crane & Walker, 1984), at least to some extent. When GM and conventional varieties of these crops coexist in the agricultural landscape, field-to-field cross-pollination by bees is a route by which GM genes could potentially be introduced into conventional crops.

Additional concern arises from a new generation of GM crops emerging from the plant biotechnology industry: plant molecular farming (Ma et al., 2005). Plant molecular farming is the cultivation of plants for the production of biomolecules useful to industry, medicine, or science. Examples of such biomolecules include vaccines, pharmaceuticals, or bioplastics (Horn et al., 2004). The advent of plant molecular farming raises public concern over the accidental incorporation of these biomolecules into human foodstuffs. Some of the plants being developed for plant molecular farming are bee-pollinated, namely safflower, melon, tomato, and tobacco (Horn et al., 2004).

Bees are potentially important pollinators in agricultural landscapes and can pose a risk to GM confinement, but is it necessary to study them specifically? For some crops, such as canola, the accumulated results of many field trials (Damgaard & Kjellsson, 2005) have shown that the spread of GM genes from a point source is likely to be widespread, but at a low level. Furthermore, these field trials also provide the basis for recommending GM confinement measures, such as separation distances between GM and conventional fields, that should ensure subthreshold GM presence in the yield from the conventional crop (Ingram, 2000). However, the high cost and effort of large field trials, such as those conducted in Australia (Rieger et al., 2002), means that it is not possible to conduct them in all possible conditions. Thus we have a limited ability to predict which landscape configurations are most susceptible to GM escape via pollen. Furthermore, field trials cannot, in themselves, identify management options, such as whether plant traits or bee abundances can be managed to improve the level of confinement. There is, therefore, benefit to be gained from principles that define and quantify the potential for cross-pollination by bees to cause escape of GM genes into conventional agriculture. Later, I discuss the causal mechanisms of insect-mediated gene flow and show how they can be modeled to theoretically assess and predict likely levels of pollinator-mediated field-to-field gene flow in a bee-pollinated crop.

Cross-pollination by bees can enable transgenes to escape from GM agriculture into various gene pools, including those of conventional crops of the same species, feral populations of the crop, or the crop's wild relatives. Understanding hybridization with wild relatives is potentially very important in managing GM confinement (Chapman & Burke, 2006; Ellstrand et al., 1999), but it is governed by many factors other than the level of cross-pollination, which is the main concern of this chapter. For example, levels of hybridization are also determined by the ability of heterospecific pollen to fertilize ovules and by the viability of the resultant offspring (Rieger et al., 2001). The discussion of the models that are presented herein focuses on pollinator-mediated gene flow between populations of conspecific plants, with the provision that the models could be adapted to apply to hybridization by incorporating additional factors.

The chapter comprises five main sections. In the first, I review our knowledge about the key aspects of the ecology of bees that make them potentially important pollinators in an agricultural landscape. In the second section, I consider how to quantify the importance of bee pollination to a particular crop. In the third section, I discuss the factors that influence field-to-field cross pollination and show how these can be quantified to evaluate the potential for GM escape. In the fourth section, I draw some general conclusions about the likely role of bees in confinement strategies for an insect-pollinated GM crop. Throughout, I draw examples exclusively from the social bees—bumble bees (Hymenoptera: Apidae, Bombini, *Bombus* spp.) and honey bees (Hymenoptera: Apidae, *Apis mellifera* L.)—that reflect their overriding importance in the European agricultural landscapes, my familiarity with these bees, and also my inability to find relevant information about other plant-pollinator systems. Finally, I address the future studies that are needed to remedy this bias and to ensure further acquisition of knowledge about bees as pollinators in agricultural landscapes. To this end, the chapter aims to provide a conceptual foundation that will enable researchers to identify the critical variables to measure.

The Potential of Bees as Pollinators in Agricultural Landscapes

Potentially, bees constitute a considerable pollinating force in certain agricultural landscapes in which, in some cases, they are capable of pollinating entire fields. Their potency as pollinators is determined by three factors: (1) their numerical abundance; (2) the rate at which they visit flowers; and (3) the effectiveness with which they transfer pollen during each flower visit. I briefly review these three key aspects.

Bees visit flowers to collect rewards such as nectar and pollen. They forage in agricultural fields when flowers are suitably rewarding. The mass rewards offered by the large numbers of flowers closely concentrated in agricultural fields can be lucrative enough to attract many bees (Scheffler et al., 1993). For example, in surveys associated with the farm-scale evaluations (FSE) of GM crops across the United Kingdom, bumble bees have been found in agricultural fields of *Brassica napus* L. (canola or oilseed rape) at area densities of up to 0.15 individuals/m^2 (Department for Environment, Food, and Rural Affairs [DEFRA], 2005); in a typical 5 ha field, this density implies the presence of 7,600 individual bees. Similarly, the FSE surveys found honey bees at area densities up to 0.26 individuals/m^2, which imply the presence of 13,000 honey bees per 5 ha field. However, in these surveys, bee abundances seldom reached these maxima, and mean area densities were an order of magnitude lower for both species.

Two biological features account for the ability of bumble bees and honey bees to arrive in large numbers at a large patch of flowers, such as the synchronous bloom of an agricultural field. First, the area density of colonies in the neighboring landscape may be high, thereby providing a large number of foragers. For example, there may be between 10 and 100 bumble bee colonies/km^2 in agricultural landscapes (Darvill et al., 2004), each holding up to several hundred foraging workers. Second, a sufficiently rich floral resource may draw bees from a wide area because their flight range is extensive. Bumble bees can fly at 7 m/s (Osborne et al., 1999), which enables them to traverse 1 km in 2.5 minutes. Thus only a small portion of the duration of a typical foraging trip (40–150 minutes; Cresswell et al., 2000) is taken up by a round trip from the bee's nest to a field that is 1 km distant. Long-distance foraging excursions are economically feasible in theory (Cresswell et al., 2000), and observations from various studies confirm this. For example, kilometer-scale foraging ranges have been determined by mark-recapture methods (Walther-Hellwig & Frankl, 2000), by analysis of spatial differentiation of molecular traits (Knight et al., 2005), and, for honey bees, by decoding their waggle dances (Visscher & Seeley, 1982). For bumble bees, up to one-third of worker bees forage more than 2 km from their nest (Walther-Hellwig & Frankl, 2000). For honey bees, foraging distances up to 6 km from the hive may be routine (Visscher & Seeley, 1982).

Individually, bees are typically assiduous and rapid floral foragers. For bumble bees and honey bees, a flower visit often takes only a few seconds in dense patches of flowers in agricultural fields. In suitable foraging areas, most movements between flowers are of short distance (Schmitt, 1980; Waser, 1982) and therefore rapidly achieved. In a field of *B. napus*, for example, a bumble bee normally visits a flower every 3 seconds and a honey bee visits one every 5 seconds (Hayter & Cresswell, 2006). In one agricultural field in the United Kingdom in July (Hayter & Cresswell, 2006), *B. napus* presented flowers at a density of $260/m^2$. Therefore, a single bumble bee required 780 seconds, or 13 minutes, to visit all the flowers in a square meter. In actuality, bumble bees were found at an exceptionally high density of 0.5 bees/m^2, and so, on average, each flower was visited by a bumble bee once every 26 minutes. Under agricultural conditions, a *B. napus* flower blooms for 5 days, or approximately 50 daylight hours, which means that at this very high abundance of bumble bees, a flower is expected to have received approximately 100 bumble bee visits during its blooming period.

Bee species vary in how effectively they pollinate a particular kind of flower (Motten et al., 1981). Large bees, such as bumble bees, are capable of delivering enough pollen to fertilize all the available ovules during a single visit to a cranberry flower (*Vaccinium macrocarpon* Aiton; Cane & Schiffhauer, 2003) or a *B. napus* flower (Mesquida & Renard, 1984; Cresswell, 1999). In contrast, full seed set in a cranberry flower requires three visits from honey bees or two from the solitary bee, *Megachile rotundata* Fab. (Cane & Schiffhauer, 2003). These limited data suggest that bees are highly effective pollinators and that flowers may require rather few bee visits to achieve full seed set.

In summary, bees have the potential to be important pollinators in agricultural landscapes, but this depends on their abundance, rate of flower visitation, and effectiveness in delivering pollen per visit. For a particular plant-pollinator interaction, I argue that the rate of flower visitation per pollinator and the effectiveness of these visits are likely to be fairly constant because they are set primarily by the physiological, morphological, and behavioral attributes of the pollinator and by the architecture of the flower, which all vary only in a limited range. For example, the observed rate at which bumble bees visit flowers of *B. napus* in agricultural fields tends to be similar in many studies (approximately 3 seconds per flower), probably because flower area densities are relatively constant in fields, constraining variation in flight times between flowers, and because the time taken by a bee to handle a flower is relatively insensitive to its nectar level (Cresswell, 1999). Variation in the pollination effectiveness per flower visit also is likely to be constrained by the conservative nature of the mechanical contacts between the sexual anatomy of the flower and the bee's body (Cresswell, 1998; Cresswell, 2000; Cresswell & Hoyle, 2006), although pollination can also vary somewhat with the availability of pollen on flowers (Cresswell, 1999). In contrast, densities of bees in agricultural fields have been observed to vary across at least two orders of magnitude (Hayter & Cresswell, 2006). I therefore argue that bee abundance is likely to be the key factor affecting variation among locations in the importance of pollinators to crop pollination (Morandin & Winston, 2005). Next, I consider the quantitative relationship between a bee species' abundance and its importance to an insect-pollinated plant species.

How Important Is Pollination by Bees in a Particular Crop?

A fundamental determinant of the importance of bee pollination in an agricultural field is the proportion of flowers that receive a bee visit, R, because bees can affect the paternity of seeds only in flowers that they visit and pollinate. R can be estimated from some basic, conveniently measured descriptors of the plant-pollinator system as follows. Let F denote the density of flowers per square meter, and let H denote the mean amount of time it takes a bee to visit a flower (including interflower travel) in hours. Hence, bees visit $1/H$ flowers per hour. If there is one bee per square meter, each flower receives a bee visit every $1/FH$ hours, on average. Let B denote the density of bees in units of individuals per square meter. Therefore, each flower is expected to receive B/FH bee visits per hour. If a flower blooms for a functional lifetime of L hours, then the total number of bee visits, or pollen deliveries, D, expected by each flower is given by

$$D = \frac{LB}{FH}. \tag{11.1}$$

If $D \leq 1$ (equation 11.1), then $R = D$; otherwise, $R = 1$. This relationship depicts only the average; the variance around this mean will depend on the behavioral details of flower visitation by the bees.

In an agricultural field of *B. napus* in the United Kingdom in April, the density of flowers was 1,150 flowers/m^2, the density of bumble bees was 0.002 bees/m^2, and $1/H$ was 1,200 flower visits/h (Hayter & Cresswell, 2006). Solving equation (11.1) with these values shows that the expected number of bumble bee visits received by a flower in its blooming period (50 h) was $D = R = 0.1$. In other words, only 10% of flowers received even a single bumble bee visit.

Clearly the importance of pollination by bees declines as R becomes smaller, but even if bees visit every flower and deliver pollen to each, other factors also determine the impact of bee pollination on seed paternity. First, a single bee visit may not fully fertilize the flower's complement of ovules (Motten et al., 1981). Second, pollen arrives on a flower's stigma by various routes (figure 11.1), and these modes of delivery compete to fertilize the flower's ovules (Harder & Routley, 2006; Hoyle et al., 2007). In some crops with flowers suited for animal pollination (i.e., zoophilous flowers), wind pollination can occur (Eisikowitch, 1981), but probably only at a relatively low rate, because

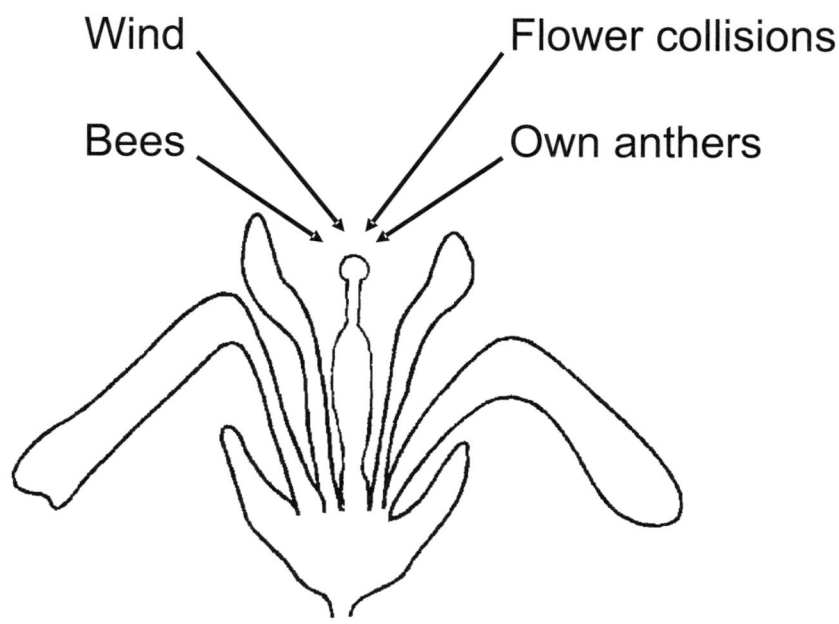

Figure 11.1 Potential modes of pollination for a zoophilous flower (i.e., a flower suited to animal pollination).

a zoophilous flower is unsuited for capturing airborne pollen (Cresswell et al., 2004). Bred for yield, it is likely that crop species are often capable of pollinating themselves autonomously, which ensures seed set even when pollinator visits are absent. Most animal-pollinated crops, for example, soybean (*Glycine max* Merr.), cotton (*Gossypium* spp.), canola (*B. napus*), and safflower (*Carthamus tinctorius* L.), can also pollinate themselves autonomously because they have hermaphrodite flowers with self-compatible pollen (Lloyd & Schoen, 1992).

Plants can pollinate themselves autonomously by various mechanisms. In soybeans, for example, pollen is released before the flower opens (Crane & Walker, 1984). In safflower, the pollen is presented by a *nudelspritze*, or noodle-squeezer, mechanism, in which the pistil acts as a piston to expel pollen that has been released inside the floral tube. The stigma is at the top of the piston, and so it becomes coated with pollen, which may later fertilize the flower's ovules (McGregor, 1976). In some wild plant species, the flower's stigma is gradually brought into contact with the stamens as the flower ages (Ruan et al., 2005). Even when special mechanisms are not evident, autonomous self-pollination can be brought about in a windblown flower when pollen is shaken from the dehiscent stamens onto the stigma (Hayter & Cresswell, 2006). When plants are closely crowded, as in many agricultural crops, collisions between windblown flowers may bring about autonomous cross-pollination (Hoyle et al., 2007).

If autonomous mechanisms deliver enough pollen to enable complete seed set, then seed yield is independent of the abundance of pollinators. Faced with this scenario, it may be tempting to conclude that bees are ineffectual pollinators and do not affect gene flow. However, consider a hypothetical example in which bee visits account, on average, for 500 of the 1,000 pollen grains that accumulate on each flower's stigma. If 500 pollen grains are sufficient to ensure full seed set, the elimination of bee visits would have no effect on yield, but the bees are nevertheless pollinating half of the ovules, assuming that pollen on the stigma fertilizes ovules in proportion to its representation on the stigma. Indeed, bee pollination could be responsible for fertilizing all the ovules if the other pollination mechanisms are late-acting, as in "delayed selfing" (Lloyd & Schoen, 1992). Thus the lack of a relationship between seed yield and bee abundance may indicate only that variation in yield is not an appropriate way to detect the pollinators' contribution to seed paternity and gene flow. Rather than analyzing the effect of bee abundance on seed yields, we need to determine the contribution made by bees to the fertilization of ovules and, hence, to gene flow. How should this be achieved?

I present two possible methods for investigating the importance of bee pollination for seed paternity in an agricultural field. Both require that the typical rate of pollen accumulation on stigmas be established (figure 11.2). To accomplish this, a cohort of flowers is marked while in bud and the time of each flower's opening is recorded. Flowers are then randomly selected for harvest at intervals, and the pollen accumulated on their stigmatic surfaces is quantified. Assuming that pollen accumulates linearly with time (figure 11.2a), let the estimated rate of total pollen accumulation be denoted by P_T grains/h.

The first method for estimating the importance of bee pollination involves combining measurements of bee abundance, activity, and effectiveness per visit to estimate the rate of pollen accumulation due to bees, denoted P_B grains/h. Various authors have proposed evaluating pollinator importance by quantifying the effectiveness of an insect species as a product of the rate at which it visits flowers and the amount of pollen it transfers per visit (Beattie, 1971; Primack & Silander, 1975), which I adapt as follows. Suppose that observations reveal that the density of a particular bee species is B individuals/m^2 and that the mean rate that individual bees visit the field's flowers is $1/H$ flowers/h. The area density of flowers in the field is F flowers/m^2. Experiments in which virgin flowers were offered to bees reveal that a single bee visit deposits G pollen grains. The estimated rate of pollen accumulation on a flower's stigma due to the bee species is given by

$$P_B = \frac{BG}{FH}. \tag{11.2}$$

The rate due to a particular bee species can then be compared with the overall, or total, rate at which pollen accumulates on the stigma, P_T. If we assume that pollen delivered by

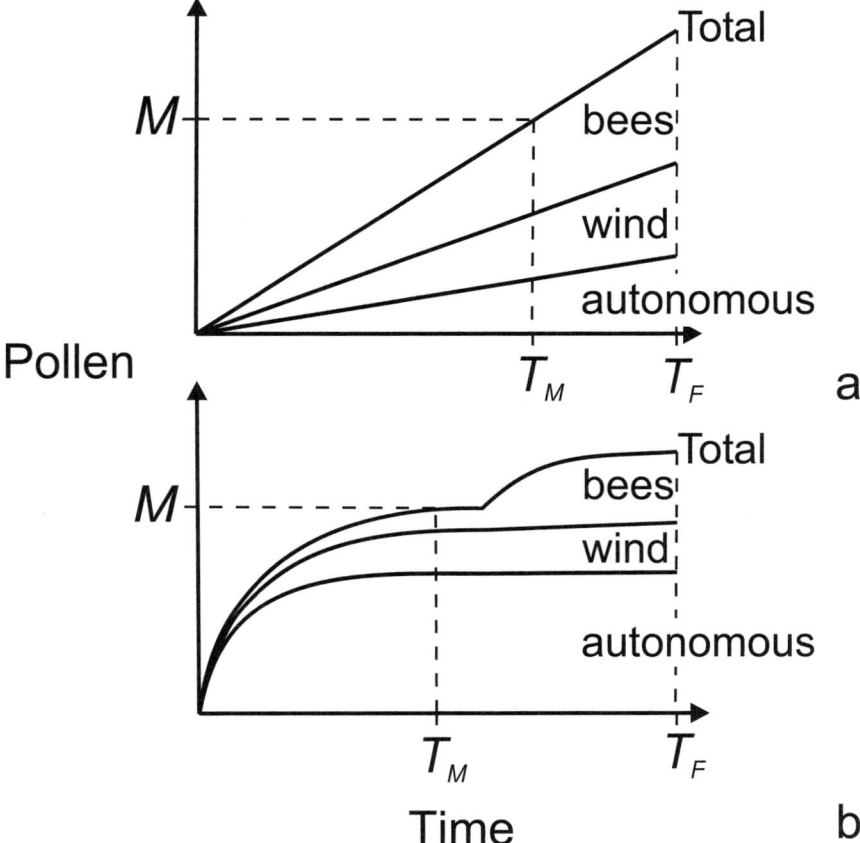

Figure 11.2 Hypothetical pollen accumulation curves. Pollen (y-axis) indicates the number of pollen grains accumulated on a flower's stigma. Time (x-axis) indicates the duration of a flower's blooming period (i.e., the flower opens at time = 0). M indicates the number of pollen grains needed for maximum seed set, T_M indicates the amount of time taken by a flower to accumulate M pollen grains, and T_F indicates the time at which a flower finishes blooming. In the upper panel (a), pollen accumulates at constant rates through each mode of pollination. In the lower panel (b), self-pollination predominates during the early life of the flower, and the rate of bee pollination increases latterly.

bees fertilizes seed in proportion to its relative abundance on stigmas, the relative magnitudes of P_B and P_T determine the bees' importance in contributing to seed paternity. Specifically, the importance of pollination by bees is indicated by the proportion of seed pollinated by these bees, I_B, which is estimated by

$$I_B = \frac{P_B}{P_T}. \tag{11.3}$$

A second method for estimating I_B involves estimating the rate of pollen accumulation by flowers from which bees are excluded. Bees can be excluded from flowers either with cages or by discouraging bee visits by making the flowers unattractive to bees through removing or discoloring the petals, which avoids the inconvenience of constructing cages and the possible artifacts in pollen accumulation rates caused in small cages when flowers collide with cage walls. The effectiveness of methods for discouraging pollinator visits must be checked, however, because bees might visit modified flowers in some instances (Pierre et al., 1996).

If the rate of pollen accumulation in flowers from which bees are excluded is denoted P_{notB} grains/h (figure 11.2), the importance of bees for the pollination of the crop is given by

$$I_B = 1 - \frac{P_{notB}}{P_T}. \tag{11.4}$$

To illustrate these approaches, consider the following data describing pollination by bumble bees in conventional agricultural fields of *B. napus* in the United Kingdom in April. Hayter and Cresswell (2006) empirically estimated the following parameter values: $B = 0.003$; $1/H = 1,200$; $F = 1,150$; and G was estimated by Cresswell (1999) as 150. Therefore, $P_B = 0.47$ (equation 11.2). Hayter and Cresswell (2006) estimated the rate of pollen accumulation by flowers as 1.4 grains/h across entire days, but assuming that pollination takes place for 10 h each day, this yields $P_T = 1.4(24/10) = 3.4$. Based on these figures, $I_B = 14\%$ (equation 11.3). Excluding pollinators in the same study system revealed that, at such low abundance, bees had no statistically detectable effect on pollen accumulation rates (Hayter & Cresswell, 2006), which implies that $P_T = P_{notB}$, and therefore $I_B = 0\%$ (equation 11.4). The latter estimate is certainly too low, however, because we observed bees pollinating flowers in the field. Nevertheless, both estimates agree that bumble bee pollination is responsible for fertilizing only a small fraction (<14%) of the seeds in this canola field, which emerges from the scarcity of bees and the high capacity of *B. napus* to pollinate itself autonomously.

The models described previously (equations 11.2–11.4) use relative contributions to pollen accumulation to estimate relative contribution to seed paternity, a method that may have pitfalls if the relative rates of the various modes of pollination vary substantially over time. All else being equal, the earliest pollen to arrive on the stigma is most likely to fertilize ovules. Pollen that arrives later may find that ovules have been preempted by pollen that has arrived even 30 minutes earlier (Snow et al., 2000). If pollen accumulation rapidly reaches the level required for full seed set, the first day, or even the first hours, after flower opening may be critical in determining the relative success of different modes of pollination. Therefore, a particular mode of pollination that predominates only during the earliest hours of stigma receptivity may accrue a fraction of ovules that is disproportionate to its relative contribution to total pollen accumulation. For example, if "prior selfing" (Lloyd & Schoen, 1992) occurs in a flower, which means that self-pollination happens while the flower is still in bud, self-pollination may dominate seed paternity even if only a small proportion of the pollen eventually accumulated arrives by this route. Similarly, if one pollinator species was particularly active when

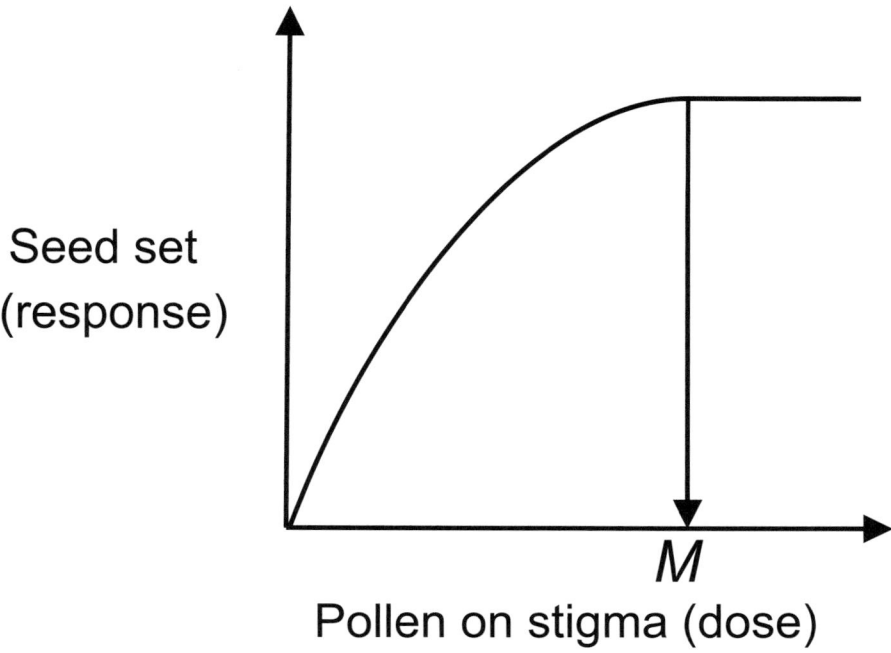

Figure 11.3 Hypothetical relationship between the number of pollen grains accumulated on a flower's stigma (dose: *x*-axis) and the number of seeds set by the flower (response: *y*-axis). M denotes the minimum number of pollen grains required for maximum seed set.

flowers first opened, perhaps in the early part of the day, its pollinating activity may make a disproportionate contribution to seed paternity. These complications could readily be revealed using a survey of pollen accumulation, as described earlier. However, the flowers must be collected frequently enough after opening to reveal any nonlinearity in pollen accumulation (figure 11.2b).

One way to begin to determine the most relevant time scale over which to study pollen accumulation is to first determine the dose-response relationship (Cane & Schiffhauer, 2003) between pollen (the dose) and seed set (the response; figure 11.3). If M denotes the minimum number of pollen grains required for maximum seed set, then the appropriate time scale over which to study pollen accumulation is T_M, the time it takes the stigma to accumulate M pollen grains.

Estimating Field-to-Field Gene Flow Mediated by Bees

Demonstrating that bees are important pollinators of a crop is necessary, but not sufficient, to implicate them as agents of genetic escape from GM crops. In this section, I analyze the determinants of levels of field-to-field cross-pollination and gene flow.

The rapid flight speed of large bees, such as bumble bees and honey bees, means that fields separated by a few tens of meters are likely to experience significant cross-pollination by bee movements (Morris et al., 1994). However, large bees also possess the necessary flight range to fly between distant fields during a single foraging bout and thereby transfer pollen across a landscape. There is, however, not enough evidence to determine whether interfield flights are sufficiently frequent to make an important contribution to gene dispersal. For foraging honey bees, most of whom are probably recruited to a rewarding site by dance communication, it seems likely that many individuals will exploit a single location before returning home. Bumble bees forage by individual initiative, but many studies of marked bumble bees have shown that individuals return to the same patches of plants over successive foraging trips (Osborne & Williams, 2001), although they may not restrict themselves to these locations. To my knowledge, only one published study demonstrates field-to-field movements, but only on succeeding days. Kreyer and colleagues (2004) found that about 5% of bumble bees marked in one field of *Phacelia tanacetifolia* were subsequently recorded foraging in another field 600 m distant. In summary, bee-mediated cross-pollination seems likely in fields separated by tens of meters, but its importance in fields separated by hundreds or thousands of meters is as yet unclear.

GM escape by field-to-field cross-pollination occurs when a bee leaves the GM field and then visits flowers in the conventional (non-GM) field, bringing with it transgenic pollen. Logically, the amount of transgenic pollen transferred to flowers in the conventional field increases with the number of bees that arrive with transgenic pollen and with the amount of pollen each bee brings. However, as a bee forages in the conventional field, its load of transgenic pollen is depleted and then increasingly diluted in the collective "pollen pool" on the stigmas of conventional flowers by nontransgenic pollen that the bee transfers among the conventional flowers themselves (figure 11.4). Therefore, the *relative* amount of transgenic pollen on the conventional field's flowers, which determines the proportion of transgenic seed, is inversely proportional to the total amount of pollen delivered by each bee during a bout of foraging in the conventional field. Equivalently, the proportion of transgenic seeds in the conventional field is determined by ratio between the number of fruits a bee fertilizes with transgenic pollen on arriving in the conventional field versus the total number of fruits the bee fertilizes during its visit to that field. This ratio quantifies the level of bee-mediated gene flow between a source population and a sink population. For plants pollinated exclusively by bees, Cresswell, Osborne, and Bell (2002) suggested that the proportion of GM-containing seed set in a conventional crop, denoted ξ_{GM}, could be modeled by

$$\xi_{GM} = \frac{E\psi}{b}, \qquad (11.5)$$

where each bee that arrives in the conventional field from the GM field fertilizes ψ fruits with GM paternity for every b conventional flowers it fertilizes, and where a proportion E of all the conventional field's bees arrive directly from the GM field.

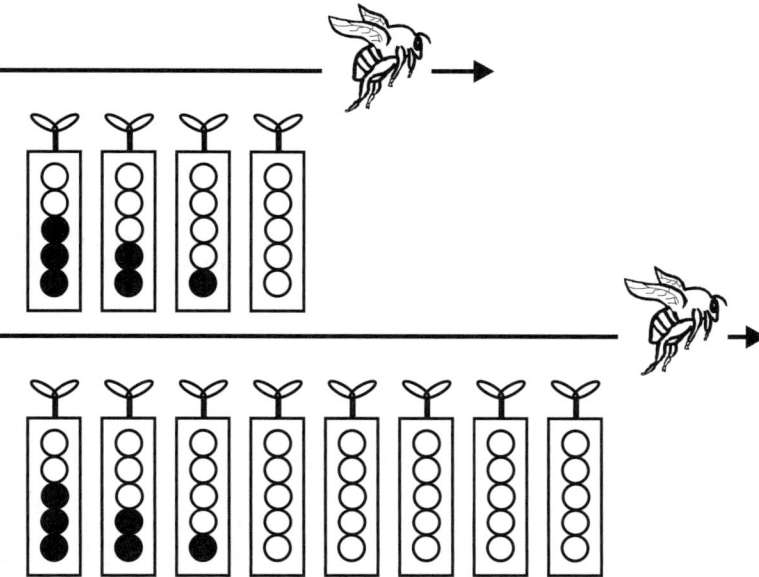

Figure 11.4 Schematic of the model in equation (11.5), which describes the consequences for seed paternity of a bee arriving and foraging in a conventional (non-GM) field carrying pollen from a GM field. The figure illustrates the influence of the number of flowers pollinated during a bout in the non-GM field, b, on the overall proportion of GM seeds, ξ. Two hypothetical foraging bouts among the field's flowers are indicated, with the shorter bout shown above. Each upright rectangle represents the fruit of a flower in the non-GM field, with its constituent seeds represented by circles. Filled circles (●) represent seeds with GM paternity, and open circles (○) represent non-GM seeds resulting from transfer of pollen within the conventional field. In equation (11.5), ψ denotes the amount of GM seed produced when a bee fertilizes b flowers, that is, the total number of fruits full of seeds with GM paternity.

To date, ψ and b have both been measured only for one plant-pollinator interaction, bumble bees and *B. napus* (Cresswell et al., 2002; Cresswell, 2005), where they were determined as $\psi \approx 1$ and $b \approx 600$. For bumble bees pollinating *B. napus*, the value of ψ is conserved across widely different schedules of pollinator visits, and it therefore appears to be a fundamental attribute of the bumble bee-*B. napus* interaction (Cresswell & Hoyle, 2006). In the most extreme case in which all arriving bees carry GM pollen, that is, $E = 1$, the solution to equation (11.5) therefore yields $\xi_{GM} = 0.2\%$. These results indicate that bee-mediated cross-pollination between agricultural fields is capable of causing GM paternity in seeds at a rate of at most a fraction of a percent. To what extent are these results likely to generalize to other bee-pollinated crops?

The outcrossing potential of the plant-pollinator interaction is set by the outcrossing parameter, ψ, which is the maximum number of fruits that the pollinator fertilizes in the conventional field with pollen from the GM field. Assuming that the pollinator has visited sufficient flowers in the GM field to become fully charged with GM pollen, the magnitude of ψ will be set by, for example, the capacity of the bee to retain pollen on its body in a position suited for retrieval by the sexual parts of flowers and by the efficacy of the pollinators' in-flight grooming. In *B. napus*, the outcrossing potential of bumble bees is $\psi \approx 1$, which means that an individual bee moving between fields arrives with the potential to fully fertilize one fruit with foreign pollen. The flower of *B. napus* pollinates itself autonomously rather slowly because the anthers dehisce only gradually (Bell & Cresswell, 1997), and so ψ is likely to be larger in *B. napus* than in plant species that rapidly bring about autonomous self-pollination, such as safflower. Moreover, the exactingly optimal fit between bumble bees' bodies and the sexual anatomy of the *B. napus* flowers that they pollinate (Cresswell, 2000) suggests that other bees are unlikely to be highly superior cross-pollinators that confer a higher value of ψ. Thus, although it is not possible to generalize with confidence about the magnitude of ψ until further studies are completed, it seems likely that agricultural crops will have $\psi \leq 1$.

The extent to which incoming GM pollen is diluted by conventional pollen is set by the dilution parameter, b, which can be estimated by the expected number of flowers that a bee visits during its bout in the conventional field. Unfortunately, the value of b has seldom been measured in agricultural fields. Rough estimates are possible based on the nectar-carrying capacity of bees in relation to the size of floral rewards. For example, in social bees the capacity of a worker's honey stomach is approximately 30 µl in honey bees (Ribbands, 1953) and approximately 80 µl in bumble bees (Heinrich, 1979). The total daily nectar production of many bee-pollinated flowers is typically less than a few microliters (Heinrich, 1979). Bees are important pollinators when their visits are frequent, in which case flowers are likely to have depleted nectar and so contain perhaps a fraction of a µl. In such a case, b will be correspondingly large, assuming that foraging bees fill their honey stomachs to capacity. This argument is supported by several observations. Bateman (1947) recorded 400 successive flights by a honey bee in a field of radish (*Raphanus sativus* L.), which implies that $b \geq 400$. For honey bees foraging on a field of white clover (*Trifolium repens*), Percival (1950, as cited in Ribbands, 1953) calculated that 585 successive visits would be required to collect a full load of pollen, and Weaver et al. (1953, as cited in Ribbands, 1953) observed a honey bee making 494 consecutive flower visits. Therefore, it is likely that only small patches of plants will be susceptible to producing seeds with high proportions of foreign paternity because of low b, that is, a small dilution effect (Cresswell & Osborne, 2004).

The likelihood that a bee that arrives at the conventional field is carrying GM pollen is given by E, the transmission parameter. The transmission parameter is estimated by the proportion of bees that arrive at the conventional field from the GM field. The value of E has so far proven impossible to quantify at a scale appropriate to an agricultural landscape. Direct observations are hampered because bees travel fast. Bumble bees, for example, fly at 7 m/s and are, therefore, difficult to follow by eye over a substantial distance. Bees do, however, take a direct route for long-distance flights, hence the term *beeline*, which connotes an undeviating route. Potentially the direction of a bee's departure from a field could reasonably serve as an indicator of the insect's intended destination (Ramsay et al., 2003). If the direction of departure from a GM field corresponds to the compass direction of a conventional agricultural field, this may signify the potential for field-to-field GM cross-pollination. However, although observations of this kind can suggest that $E > 0$, they do not define the value of E any further. Harmonic radar can trace at high resolution the movements of small numbers of bees individually as they travel over landscapes up to 500 m from the radar source (Osborne et al. 1999), but this method also does not quantify the origins of a collective mass of bees arriving at a field, and therefore is not suited to quantifying E. At present, there are no empirical estimates of E, and studies using the model of equation (11.5) have calculated only "worst case" estimates using $E = 1$ (Cresswell et al., 2002). However, it may be possible to estimate E theoretically (Hoyle & Cresswell, 2007).

Equation (11.5) must be modified if bee abundances are insufficient to ensure that every flower receives a bee visit. If only a proportion of a conventional field's flowers receive a visit by a bee (i.e., $R < 1$; see equation 11.1) then the proportion of GM seed due to bees is given by

$$\xi_{GM} = R\left(\frac{E\psi}{b}\right). \tag{11.6}$$

In an agricultural field of *B. napus* that bloomed in April in the United Kingdom, bees were scarce, such that $R = 0.1$ (Hayter & Cresswell, 2006). Assuming $E = 1$ in equation (11.6), the maximum potential for bumble bee-meditated gene flow is $\xi_{GM} = 0.02\%$. If GM pollen arrives by other means, such as wind pollination, then the overall proportion of seed with GM paternity may exceed ξ_{GM}, however. When bees are scarce (i.e., not all flowers receive a bee visit and $R < 1$), then following equation (11.1) we can write

$$\xi_{GM} = \frac{LB}{FH}\left(\frac{E\psi}{b}\right). \tag{11.7}$$

In this case, the likelihood that bees will mediate GM gene flow is inversely related to flower area density, F, because high flower density decreases the chances of flowers receiving a bee visit, and proportional to flower lifespan, L, because flower longevity prolongs exposure to bee visits.

Implications for the Confinement of Bee-Pollinated GM Crops

The potential importance of bees as a threat to the confinement of GM crops should be evaluated on a case-by-case basis, but I argue that in many instances bees are unlikely to cause high levels of genetic escape into large agricultural fields of conspecific conventional crop varieties for three reasons: (1) it is possible for the millions of flowers in a large agricultural field to saturate the available bees; (2) even when visits to flowers by bees are common, the importance of pollination by bees is reduced by competing modes of pollination that are likely to be active in crops bred for yield assurance, for example, autonomous self-pollination; and (3) even if bees travel frequently between GM and conventional fields, field-to-field cross-pollination by bees is likely to produce only low levels of gene flow provided that, on average, bees visit a large number of flowers during each visit to the field.

Recommendations for improving confinement in bee-pollinated GM crops based on these conclusions are as follows. Flowers that pollinate themselves autonomously reduce the proportion of the seed that is fertilized by bees. Floral traits that improve the rapidity of autonomous self-pollination could be targeted by plant breeders to further insulate seed of conventional agricultural varieties against fertilization by foreign pollen. When bees are scarce, the importance of pollination by bees is reduced as the density of flowers increases and flower longevity decreases. Farmers can control flower density by modifying planting density, and plant breeders can adjust floral longevity. Two other possible strategies related to bee pollination could contribute to GM confinement, but our current level of knowledge is insufficient to clarify their worth.

First, it is theoretically possible to adjust a crop's rate of nectar production, because it is a heritable trait. If reduced nectar production by the flowers in a conventional field led to decreased amounts of nectar in flowers, nectar-collecting bees would have to visit more flowers in a field to reach capacity, which would increase b and thereby reduce ξ, the proportion of the seed yield that contains foreign paternity (equation 11.5). However, bees encountering low nectar levels may instead be more likely to quit the field and explore elsewhere, thereby decreasing b and increasing ξ. The likely effect on ξ of increasing nectar production in the conventional field is equally uncertain. Higher nectar rewards may cause bees to visit fewer flowers before reaching capacity, thereby decreasing b and increasing ξ. Alternatively, higher nectar levels in the conventional field may promote foraging site fidelity by the bees, which would decrease E and thereby reduce ξ. Thus it is not possible to evaluate further these opportunities until more is known about the relationship between floral rewards and landscape-scale movements by bees.

Second, it is theoretically possible to alter the composition of the pollinator fauna. Most feasibly, the abundance of honey bees in the vicinity of a non-GM field could be increased by introducing domesticated hives. If honey bees showed site fidelity to the non-GM field for entire foraging bouts, within-field pollination would be augmented, thereby insulating the field against foreign pollen. However, the site fidelity of honey bees is not fully determined, and so the effect of this manipulation is currently unclear. Moreover, it is possible that pollen transfer in the honey bee hive may further spread GM pollen (Ramsay et al., 1999).

Overall, however, ecologically based strategies for improving GM confinement are likely to be less important than strategies based in the biotechnology itself (Chapman & Burke, 2006).

The Agenda for Future Research

The theoretical models presented in this chapter are intended to promote progress in understanding the importance of bees as pollinators in the agricultural landscape. In particular, the parameters of the models are intended to begin to expose the fundamental influences on pollination and plant gene flow in plant-pollinator interactions. Furthermore, some of the parameters are convenient to measure. For example, the area densities of flowers and bees, floral longevity, and bee foraging rates are simple to estimate in the field and potentially highly informative about the importance of bees as pollinators (equation 11.1). Despite the ease with which these data could be quantified, it is difficult to find studies in which they are comprehensively reported. Additionally, we need more studies of the capacities of single bee visits to deliver pollen and thereby fertilize seed. Cane and Schiffhauer's (2003) study of bee pollination in cranberry is exemplary in this respect, and more like it must be made in other systems. However, a full understanding of the plant-pollinator interaction requires cooperation between both entomologists and botanists. The quantitative study of competing modes of pollination has been neglected for too long, and proper recognition must be given to the role of the plant's breeding system and sexual morphology (equations 11.3 and 11.4). Finally, although accurate theoretical prediction of field-to-field gene flow is now possible (equation 11.6), the technical difficulties of measuring field-to-field movements by bees need to be overcome, and ingenious scientists must give this some thought.

References

Barber, S. (1999). Transgenic plants: Field testing and commercialization including a consideration of novel herbicide resistant rape (*Brassica napus* L.). In P. J. Lutman (Ed.), *Gene flow and agriculture* (3–12). Farnham, Surrey, UK: British Crop Protection Council.

Bateman, A. J. (1947). Contamination of seed crops: I. Insect pollination. *Journal of Genetics, 48*, 257–275.

Beattie, A. J. (1971). Pollination mechanisms in *Viola*. *New Phytologist, 70*, 343–360.

Belcher, K., Nolan, J., & Phillips, P. W. B. (2005). Genetically modified crops and agricultural landscapes: Spatial patterns of contamination. *Ecological Economics, 53*, 387–401.

Bell, S. A., & Cresswell, J. E. (1997). The phenology of gender in homogamous flowers: Temporal change in the residual sex function of flowers of oil-seed rape *Brassica napus*. *Functional Ecology, 12*, 298–306.

Cane, J. H., & Schiffhauer, D. (2003). Dose-response relationships between pollination and fruiting refine pollinator comparisons for cranberry (*Vaccinium macrocarpon* [Ericaceae]). *American Journal of Botany, 90*, 1425–1432.

Chapman, M. A., & Burke, J. M. (2006). Letting the gene out of the bottle: The population genetics of genetically modified crops. *New Phytologist, 170*, 429–443.

Colwell, R. K., Norse, E. A., Pimental, D., Sharples, F. E., & Simberloff, D. (1985). Genetic engineering in agriculture. *Science, 229*, 111–112.

Crane, E., & Walker, P. (1984). *Pollination directory of world crops*. London: International Bee Research Association.

Cresswell, J. E. (1998). Stabilising selection and the structural variability of flowers within species. *Annals of Botany, 81*, 463–473.

———. (1999). The influence of nectar and pollen availability on pollen transfer by individual flowers of oil-seed rape (*Brassica napus*) when pollinated by bumblebees (*Bombus lapidarius*). *Journal of Ecology, 87*, 670–677.

———. (2000). Manipulation of female architecture in flowers reveals a narrow optimum for pollen deposition. *Ecology, 81*, 3244–3249.

———. (2005). Accurate theoretical prediction of pollinator-mediated gene dispersal. *Ecology, 86*, 574–578.

Cresswell, J. E., Davies, T. W., Patrick, M. A., Russell, F., Pennel, C., Vicot, M., et al. (2004). The aerodynamics of wind pollination in a zoophilous flower. *Functional Ecology, 18*, 861–866.

Cresswell, J. E., & Hoyle, M. (2006). A mathematical method for estimating patterns of flower-to-flower gene dispersal from a simple field experiment. *Functional Ecology, 20*, 245–251.

Cresswell, J. E., & Osborne, J. L. (2004). The effect of patch size and separateness on bumblebee foraging in oilseed rape (*Brassica napus*): Implications for gene flow. *Journal of Applied Ecology, 41*, 539–546.

Cresswell, J. E., Osborne, J. L., & Bell, S. A. (2002). A model of pollinator-mediated gene flow between plant populations with numerical solutions for bumblebees pollinating oilseed rape. *Oikos, 98*, 375–384.

Cresswell, J. E., Osborne, J. L., & Goulson, D. (2000). An economic model of the limits to foraging range in central place foragers with numerical solutions for bumblebees. *Ecological Entomology, 25*, 249–255.

Damgaard, C., & Kjellsson, G. (2005). Gene flow of oilseed rape (*Brassica napus*) according to isolation distance and buffer zone. *Agriculture Ecosystems and Environment, 108*, 291–301.

Darvill, B., Knight, M. E., & Goulson, D. (2004). Use of genetic markers to quantify bumblebee foraging range and nest density. *Oikos, 107*, 471–478.

Department for Environment, Food and Rural Affairs. (2005). *Farm scale evaluations of genetically modified herbicide tolerant crops*. http://www.farmscale.org.uk.

Eisikowitch, D. (1981). Some aspects of pollination of oil-seed rape (*Brassica napus* L.). *Journal of Agricultural Science, Cambridge, 96*, 321–326.

Ellstrand, N. C., Prentice, H. C., & Hancock, J. F. (1999). Gene flow and introgression from domesticated plants into their wild relatives. *Annual Review of Ecology and Systematics, 30*, 539–563.

Harder, L. D., & Routley, M. B. (2006). Pollen and ovule fates and reproductive performance by flowering plants. In L. Harder & S. Barrett (Eds.), *Ecology and evolution of flowers* (61–80). Oxford, UK: Oxford University Press.

Hayter, K., & Cresswell, J. (2006). The influence of pollinator abundance on the dynamics and efficiency of pollination in arable *Brassica napus*: Implications for landscape-scale gene dispersal. *Journal of Applied Ecology, 43*, 1196–1202.

Heinrich, B. (1979). *Bumblebee economics*. Cambridge, MA: Harvard University Press.

Horn, M. E., Woodard, S. L., & Howard, J. A. (2004). Plant molecular farming: Systems and products. *Plant Cell Reports, 22*, 711–720.

Hoyle, M., & Cresswell, J. E. (2007). A model of patch-to-patch forager movement with application to pollen-mediated gene flow. *Journal of Theoretical Biology 248*, 154–163.

Hoyle, M., Hayter, K. E., & Cresswell, J. E. (2007). Effect of pollinator abundance on self-fertilization and gene flow: Application to GM canola (*Brassica napus*). *Ecological Applications, 17*, 2123–2135.

Ingram, J. (2000). The separation distances required to ensure cross-pollination is below specified limits in non-seed crops of sugar beet, maize and oilseed rape. *Plant Varieties and Seeds, 13*, 181–199.

James, C. (2004). *Global status of commercialized biotech/GM crops* (ISAAA Briefs No. 32). Ithaca, NY: International Service for the Acquisition of Agri-biotech Applications.

Knight, M. E., Martin, A. P., Bishop, S., Osborne, J. L., Hale, R. J., Sanderson, A., & Goulson, D. (2005). An interspecific comparison of foraging range and nest density of four bumblebee (*Bombus*) species. *Molecular Ecology, 14*, 1811–1820.

Kreyer, D., Oed, A., Walther-Hellwig, K., & Frankl, R. (2004). Are forests potential landscape barriers for foraging bumblebees? Landscape scale experiments with *Bombus terrestris* Agg. and *Bombus pascuorum* (Hymenoptera, Apidae). *Biological Conservation, 116*, 111–118.

Lloyd, D. G., & Schoen, D. J. (1992). Self- and cross-fertilization in plants: I. Functional dimensions. *International Journal of Plant Sciences, 153*, 358–369.

Lutman, P. J. W. (1999). *Gene flow and agriculture*. Farnham, Surrey, UK: British Crop Research Council.

Ma, J. K. C., Chikwarnba, R., Sparrow, P., Fischer, R., Mahoney, R., & Twyman, R. M. (2005). Plant-derived pharmaceuticals: The road forward. *Trends in Plant Science, 10*, 580–585.

McGregor, S. (1976). *Insect pollination of cultivated crops: Agricultural handbook 496*. Washington, DC: U.S. Department of Agriculture.

Mesquida, J., & Renard, M. (1984). Etude des quantités de pollen déposées sur les stigmates dans différents conditions de pollinisation; influence sur la production de graines chez le colza d'hiver male-fertile. *Proceedings of the Vth International Symposium on Pollination* (Vol. 19, pp. 351–356).Versailles: French National Institute for Agricultural Research.

Morandin, L. A., & Winston, M. L. (2005). Wild bee abundance and seed production in conventional, organic, and genetically modified canola. *Ecological Applications, 15*, 871–881.

Morris, W. F., Kareiva, P. M., & Raymer, P. L. (1994). Do barren zones and pollen traps reduce gene escape from transgenic crops? *Ecological Applications, 4*, 157–165.

Motten, A. F., Campbell, D. R., Alexander, D. E., & Miller, H. L. (1981). Pollination effectiveness of specialist and generalist visitors to a North Carolina population of *Claytonia virginica*. *Ecology, 62*, 1278–1287.

Osborne, J. L., Clark. S. J., Morris, R. J., Williams, I. H., Riley, J. R., Smith, A. D., et al. (1999). A landscape-scale study of bumble bee foraging range and constancy, using harmonic radar. *Journal of Applied Ecology, 36*, 519–533.

Osborne, J. L., & Williams, I. H. (2001). Site constancy of bumblebees in an experimentally patchy habitat. *Agriculture, Ecosystems and Environment, 83*, 129–141.

Pierre, J., Pierre. J. S., Marilleau, R., PhamDelegue, M. H., Tanguy, X., & Renard, M. (1996). Influence of the apetalous character in rape (*Brassica napus*) on the foraging behavior of honeybees (*Apis mellifera*). *Plant Breeding, 115*, 484–487.

Poppy, G., & Wilkinson, M. (Eds.). (2005). *Gene flow from GM plants*. Oxford, UK: Blackwell.

Primack, R. B., & Silander, J. A. (1975). Measuring the relative importance of different pollinators to plants. *Nature, 255,* 143–144.

Ramsay, G., Thompson, C., & Squire, G. (2003). *Quantifying landscape-scale gene flow in oilseed rape.* London: Department for Environment, Food and Rural Affairs.

Ramsay, G., Thompson, C. E., Neilson, S., & Mackay, G. R. (1999). Honeybees as vectors of GM oilseed rape pollen. In P. J. W. Lutman (Ed.), *Gene flow and agriculture: Relevance for transgenic crops* (57–64). Nottingham, UK: Major Design and Production.

Ribbands, C. (1953). *The behaviour and social life of honeybees.* London: Bee Research Association.

Rieger, M. A., Lamond, M., Preston, C., Powles, S. B., & Roush, R. T. (2002). Pollen-mediated movement of herbicide resistance between commercial canola fields. *Science, 296,* 2386–2388.

Rieger, M. A., Potter, T. D., Preston, C., & Powles, S. B. (2001). Hybridisation between *Brassica napus* L. and *Raphanus raphinistrum* L. under agronomic field conditions. *Theoretical and Applied Genetics, 103,* 555–560.

Ruan, C. J., Qin, P., & Xi, Y. G. (2005). Floral traits and pollination modes in *Kosteletzkya virginica* (Malvaceae). *Belgian Journal of Botany, 138,* 39–46.

Scheffler, J. A., Parkinson, R., & Dale, P. J. (1993). Frequency and distance of pollen dispersal from transgenic oilseed rape (*Brassica napus*). *Transgenic Research, 2,* 356–364.

Schmitt, J. (1980). Pollinator foraging behaviour and gene dispersal in *Senecio* (Compositae). *Evolution, 34,* 934–943.

Snow, A. A., Spira, T. P., & Liu, H. (2000). Effects of sequential pollination on the success of "fast" and "slow" pollen donors in *Hibiscus moscheutos* (Malvaceae). *American Journal of Botany, 87,* 1656–1659.

Visscher, P. K., & Seeley, T. D. (1982). Foraging strategy of honeybee colonies in a temperate deciduous forest. *Ecology, 63,* 1790–1801.

Walther-Hellwig, K., & Frankl, R. (2000). Foraging habitats and foraging distances of bumblebees, *Bombus* spp. (Hym., Apidae), in an agricultural landscape. *Journal of Applied Entomology, 124,* 299–306.

Waser, N. M. (1982). A comparison of distances flown by different visitors to flowers of the same species. *Oecologia, 55,* 251–257.

Weekes, R., Deppe, C., Allnutt, T., Boffey, C., Morgan, D., Morgan, S., et al. (2005). Crop-to-crop gene flow using farm scale sites of oilseed rape (*Brassica napus*) in the UK. *Transgenic Research, 14,* 749–759.

12 Genetically Modified Crops

Effects on Bees and Pollination

Lora A. Morandin

Introduction

Genetic engineering of crop plants has opened up profoundly new avenues of species alteration beyond what was possible with traditional plant breeding or mutagenesis. Essentially, the entire community of life has become a source of new genes for crops (Chrispeels & Sadava, 2003). Accompanying the creation of novel crop varieties is the potential for environmental impact, beneficial or harmful.

Commercialization of genetically modified (GM) crops has raised considerable concern about negative impacts they may have on nontarget organisms. Bees are nontarget organisms that have intrinsic environmental value and that also directly benefit crop production through their role as pollinators, making the question of how GM crops affect bees an economic, as well as a conservation, issue. Genetic modification does not refer to one or even a few types of alterations of plants, but rather to a process that is used to insert novel genes that can be as widely different from each other as they may be from the environmental impacts they cause. With this in mind, it becomes clear why there is no simple answer to the question of potential harm to bees or other organisms from GM crops.

Crops have been systematically improved through plant breeding for over 150 years. Common goals of crop breeding programs are improved tolerance to pests and abiotic stresses, improved productivity, and improved processing and nutritional characteristics (Beversdorf, 1993). Although world hunger may be more related to food distribution and unequal access than to insufficient food production per se (Matson et al., 1997), further agricultural expansion, cultivar improvement, and intensification are inevitable (Tilman et al., 2001), with the world's population expected to grow from 6.5 billion at present to over 9 billion by 2050 (U.S. Census Bureau, 2008). Despite controversy, cultivation of GM crops is an integral present and future component of agricultural development.

The DNA of all organisms is fundamentally identical, so that genes and parts of genes can be exchanged. In the 1970s the first recombinant organisms were created, and by the mid-1980s field trials had begun with GM crop plants. As of 2004, 17 countries around the world were growing at least one transgenic crop. The leaders in GM crop production by area grown (millions of ha; percent of total) are the United States (47.6; 59%), Argentina (16.2; 20%), Canada (5.4; 6%), Brazil (5; 6%), and China (3.7; 5%). The rate of growth of the GM crop industry has increased since commercialization of GM crops in 1996, and between 2003 and 2004 there was a 20% increase in area used for genetically modified crop production. Twenty-nine percent of corn, cotton, soybean, and canola are now genetically modified globally. Herbicide tolerance has been the dominant trait introduced into crop species (72% of GM acreage), followed by insect resistance (19% of GM acreage; James, 2005). Currently, the unofficial moratorium on GM crops maintained by the European Union (EU) since 1998, which prevents farmers from growing GM crops, is weakening. Field trials of beet, oilseed rape, and maize have been conducted in the past few years in the United Kingdom, and Britain approved GM herbicide-tolerant maize in 2004 for commercial production.

Pollination by insects, primarily bees, improves crop yield in approximately 80% of the roughly 100 crop plants that feed the world (Prescott-Allen & Prescott-Allen, 1990; Ingram et al., 1996), and it is estimated that 30% of food in developed countries results from bee pollination. Some crop species rely completely on bee-mediated pollen transfer, whereas others are self-fertile yet have improved seed or fruit production with bee pollination.

All bees require nectar and pollen sources throughout their adult life for their own energetic needs and to provision their offspring. Honey bees are the most economically important managed pollinators, and studies of effects of conventional and transgenic pesticidal proteins largely have focused on impacts to this one species of bees. Yet bees (Superfamily: Apoidea) are a very diverse group, with 20,000 to 30,000 species from seven families worldwide, ranging from solitary to colonial to primitively social species to the highly social bees (Michener et al., 1994). Pesticides and agricultural practices could have different impacts on managed honey bees and other bees. For example, care can be taken to protect managed honey bee colonies from pesticide sprays by closing colonies or moving colonies to different crops during spraying. Size, foraging range, and physiological, social, and behavioral differences among wild bee species also play a role in determining what impacts GM technology may have. Testing at least a few bee species from genera other than *Apis* would provide some knowledge of the sensitivity of other bees to commonly used GM crops. I distinguish between effects on managed honey bees and wild bees when discussing GM technology. However, little data are available on impacts of GM agriculture on wild bees, and I, therefore, cannot provide much detail on how products and practices affect different wild bee species.

Unfortunately, new technologies often are adopted before environmental impacts are reasonably well understood. Neglecting environmental impacts of agriculture threatens to disrupt ecosystem services such as pollination, possibly reducing or eliminating any yield benefit from agricultural intensification and expansion. Two fundamental questions regarding bees and genetically modified crops are: (1) do GM crops have an impact on bees and bee populations either negatively or positively and (2) if GM crops do have an impact on bees, will this affect crop pollination and yield?

In this chapter, I explore some scientific evidence for a possible impact, or lack of impact, of this relatively new technology on bees and crop pollination. I examine the state of our current understanding and areas in which information is scarce on the interaction between bees and GM crops. And I explore the questions that should be asked and suggest research that should be done in order to mitigate the negative effects of genetic modification of crop plants on bees.

Potential Effects of GM Crops on Bees

Two main types of impacts that GM crops could have on bees are direct toxic effects and indirect agroecosystem effects. Toxicity tests usually begin with controlled laboratory experiments using the product of genetic modification, a purified transgene protein. Laboratory experiments sometimes are followed by field trials on the health of bee colonies (in the case of honey bees) that forage on plots of the transgenic crop. Most commonly, toxicity is assessed in laboratory experiments that evaluate bee mortality rates, but bees also need to perform complex behaviors to collect pollen and nectar and to provision their offspring, stimulating studies on potential effects of toxins on learning and foraging behavior of exposed bees.

Genetic modification of plants could have indirect effects on bee communities by altering the environment that bees experience. Direct effects involve toxicity of the protein product expressed by the inserted gene, whereas indirect effects involve unintentional alteration of the modified plant or differences in agricultural practices associated with the GM cultivar. Field studies using realistically sized trial plots or commercial fields need to be conducted to assess indirect effects of GM crops on bees, but few such studies have been conducted. To test for direct toxicity of pesticidal proteins on bees, the United States Environmental Protection Agency (EPA) requires that the modified crops be tested for safety on honey bees using partial-field tests comparing the health of colonies exposed to GM plants and equivalent non-GM plants (Z. Vaituzis, EPA, personal communication, March 8, 2006). The EU follows international regulations outlined in the Cartagena Protocol on Biosafety (2000), which states that risk assessment be conducted in order "to identify and evaluate the potential adverse effects of living modified organisms on the conservation and sustainable use of biological diversity in the likely receiving environment" (p. 28).

Currently, no set guidelines are in place for direct and indirect testing on bees. In the United States, environmental testing of GM plants, other than testing for direct toxicity of transgenic pesticidal proteins, is regulated by the Animal and Plant Health Inspection Service (APHIS) of the U. S. Department of Agriculture (USDA). As part of the approval process, APHIS requires the applicant to adequately prove, either through literature or experiments, that the transformed crop will not harm bees (R. Rose, APHIS, personal communication, March 2006). However, when APHIS determines that direct toxic effects are unlikely to occur, they do not assess potential for indirect effects.

Gene insertions that confer insect resistance to crops have the potential to cause harm to beneficial insects, such as bees. Bees consume primarily nectar and pollen. Nectar contains insignificant amounts of protein and is unlikely to contain transgenic protein products, but pollen is 8–40% protein and is the most likely route of transgene protein exposure to bees.

Transgene products from herbicide-tolerant crops are not likely to harm nontarget beneficial insects such as bees, and this supposition has been supported by a number of studies using direct protein feeding, semifield, and field experiments (Pierre et al., 2003; Huang et al., 2004). GM crops with insect-resistant transgene products are more likely to be harmful to beneficial insects because of a relatively similar physiology among insects.

A potentially novel problem that is presented with the use of GM insect-resistant crops is that bees could be constantly exposed to pesticides throughout crop bloom. Applied pesticides break down in the environment at varying rates, and application can be restricted to times at which a crop is not in bloom or at night, minimizing pollinator exposure. Duration and level of exposure is an important component of the risk to bees, and relative risk among insect control strategies must be assessed in a way that mimics potential environmental exposure.

Although it is obvious that lethal effects of pesticides on bees would cause harm to populations, sublethal effects also could harm bees. Bees must perform complex behaviors that combine motor and learning skills in order to locate food sources and efficiently provision their offspring. Toxins that disrupt a bee's ability to perform these tasks through disruption of physiology and/or behavior could result in decreased reproduction, with negative long-term population implications. Sublethal effects of pesticides on bees are not as obvious as lethal effects and thus may be more difficult to assess. However, if overlooked, sublethal effects could cause unexpected impacts to bee populations that are possibly greater than more well-understood lethal effects.

Direct Toxic Effects of Pesticidal GM Plant

Bt Proteins

More than 99% of the commercialized GM insect-resistant crops have been transformed with genes coding for crystalline (Cry) proteins from the soil bacterium *Bacillus thurigiensis* (Bt; James, 2005). Bt is a gram-positive bacterium that produces crystalline proteins during sporulation (Simpson et al., 1997), and various strains show insecticidal activity against Lepidoptera, Diptera, Colepotera, and Hymenoptera (Hofmann & Luthy, 1986; Benz & Joeressen, 1994). The crystalline proteins are dissolved in the midgut of some insect species, releasing one or more potential toxins. The toxins may be cleaved by gut proteases, resulting in active toxins that bind to specific epithelial cells, causing pore formation and resulting in an almost immediate cessation of feeding, followed by death (Gill et al., 1992). Specificity of Bt toxins may result from specific binding affinity of certain Bt toxins to gut tissues of species or groups of species; however, the mode and degree of specificity of Bt toxins are poorly understood (Simpson et al., 1997).

Bt formulations have been used in surface application form as a low-risk and organic insect control treatment since the 1920s, and there have been no observed negative impacts on bees. In the late 1980s and early 1990s, genes coding for the active crystalline proteins were inserted into plants, and the first GM-Bt crop plants were grown commercially in 1996. Traditional applications of Bt contain whole bacteria and spores, and the proteins have to be broken down in the gut of insects to produce an active toxin. However, transgenic Bt plants contain genes coding for the active toxin, possibly resulting in negative impacts from transgenic lines that are not found with traditional Bt applications. Depending on the promoter used and the transformation line, Cry and other transgene protein products may be expressed in pollen (Wilkinson et al., 1997) of transformed plants. Cotton plants containing the Cry1Ac gene from Bt (BollGard, Monsanto, St. Louis, MO) express the protein in pollen at a concentration of 11.5 ng/g fresh weight, whereas concentrations in nectar are below detectable levels of 1.6 ng/g (U.S. Environmental Protection Agency, 2001). Expression of Cry1Ab within pollen has been found to vary greatly depending on the promoter gene involved, ranging from 1.7 to 0.09 µg/g pollen (Sears et al., 2001). Unfortunately, few results are available on expression levels of Bt or other transgene products in pollen, making it difficult for researchers to design laboratory experiments testing effects of realistic exposure on beneficial insects.

An early experiment of Cry proteins on monarch butterfly larvae indicated that there could be some harm to nontarget insects (Losey et al., 1999), sparking an eruption of controversy over the use of Bt crops and a wave of new studies. Prior to the Losey et al. (1999) study, the EPA had conducted risk assessment studies of Bt toxins on a wide range of insects, birds, and mammals and concluded that they could foresee no substantial risk to humans, nontarget organisms, or the environment (U.S. Environmental Protection Agency, 2000). Some researchers had concerns before and after publication of the Losey study that the laboratory methods used were not adequate for making conclusions regarding possible field effects. Yet a public perception developed that monarch butterfly populations would be harmed by the use of Bt crops. A substantial body of peer-reviewed literature now exists, primarily resulting from collaborative studies between Canadian and U.S. scientists, that found that levels of Bt in pollen had negligible impact on monarch butterfly populations (Sears et al., 2001).

Research on other nontarget insects, particularly honey bees, also has been pursued. Malone and Pham-Delegue (2001) summarized findings on effects of Bt on larval and adult honey bee food consumption, flight activity, and longevity. They found no evidence that any of the purified Bt proteins tested (lepidopteran-active Cry1Ac, Cry1Ab, Cry1Ba, and Cry9C and coleopteran-active Cry3A and Cry3B) would have a negative impact on honey bees, even at extremely elevated doses (see Malone & Pham-Delegue, 2001, for review), although hymenopteran-active strains (Benz & Joeressen, 1994) have not been tested on bees. More recently, Hanley et al. (2003) assessed honey bee larvae fed Cry1Ab and Cry1F proteins expressed in corn pollen and reached a similar conclusion of no negative impact. Malone et al. (2004) assessed gland development of newly emerged adult honey bees fed Cry1Ba and found no negative impact, and Liu et al. (2005) tested Cry1Ac cotton pollen on adult honey bees and also found no negative impacts.

Few studies are available on the effects of Bt toxins on any bees other than honey bees. However, because of behavioral and morphological variation among bee species, data from one species is not necessarily transferable to another. Morandin and Winston (2003) tested the effects of Cry1Ac on bumble bee adult survival, weight, and longevity, larval survival, and on the foraging ability of adult bees that had fed on the toxin during larval development and as adults. We found no difference in survival, weight, longevity, or foraging ability of bees fed Cry1Ac and control bees, and commercial use of Cry1Ac is not expected to have a negative impact on bumble bees.

Based on the number of studies, the range of testing methods, and the range of Bt proteins assessed, it can be concluded that Bt proteins expressed in GM crops will likely not harm honey bees or bumble bees and that honey bees and bumble bees are not negatively affected by purified Bt proteins at levels approximating concentrations found in transgenic pollen or by GM crops expressing proteins derived from Bt.

Protease Inhibitors

Currently, GM-Bt crops are the only commercially grown GM insect-resistant crops, but many types of transgenic proteins are being developed and tested for insect control. Protease inhibitors are a group of insecticidal proteins that can be isolated from plants, animals, and microorganisms. Protease inhibitors bind to digestive proteinases, disrupting protein digestion, resulting in slowed growth and/or insect death. They have activity against Coleoptera, Lepidoptera, and Orthoptera (Malone et al., 2004), and plants expressing protease inhibitors have been shown to protect crops from insect pests (see Malone & Pham-Delegue, 2001, and references therein). Protease inhibitors are not as specific as Bt toxins, and honey bees, bumble bees, and possibly other bees have digestive serine proteinases in their gut, making them susceptible to these pesticides (Malone et al., 2004; Dechaume-Moncharmont et al., 2005).

Malone and Pham-Delegue (2001) reviewed findings on the effects of various protease inhibitors on honey bees and bumble bees and found that some show negative impact and others show no impact. They concluded that ingestion of high doses of protease inhibitors by bees may reduce gut protease activities, decreasing longevity. They emphasize that a case-by-case approach needs to be applied to each new transgenic protein product. Recently, other studies have found differing effects of protease inhibitors on bees. For example, the Kunitz Soybean Trypsin Inhibitor (SBTI), when fed to honey bee larvae, caused significant larval mortality, slowed juvenile development, and reduced adult body mass (Brodsgaard et al., 2003). The development of honey bee hypopharyngeal glands was not affected when honey bee larvae were fed the protease inhibitor aprotinin (Malone et al., 2004), but development was negatively affected and larval survival was lower in bees fed SBTI (Sagili et al., 2005). In combination, these studies highlight the need to examine each new protease inhibitor transgene on multiple measures of bee health.

Other Insecticidal Proteins

Many studies on pest-resistant GM plants have examined Bt and protease inhibitors; however, other transgenic plants are being tested that express biotin-binding proteins, chitinases, gluconases, lectins, and spider venom (Malone & Pham-Delegue, 2001). Although initial studies indicate that chitinases and glucanase proteins are unlikely to harm honey bees (Picard-Nizou et al., 1997) and that chitinases are not harmful to bumble bees (Morandin & Winston, 2003), few doses have been tested. Studies are lacking on effects of other potential transgene proteins.

Positive Impacts of GM Pest-Resistant Crops

So far I have focused on potential negative impacts that adoption of GM pest control technology could have on bees, but there is also the possibility that this new technology could have positive impacts on bees. Reduced use of chemical pesticides is one such positive impact.

Since 1945 synthetic pesticide use has increased tenfold, and currently about 2.5 million tons of pesticides are applied per year worldwide (Paoletti & Pimentel, 2000). Even with ever increasing use of pesticides, it is estimated that 40% of crop yields are lost to pests before harvest; yet without pesticides, crop losses could increase another 30%. The broad-spectrum synthetic pesticides that are most commonly used can be toxic to bees, as well as to other wildlife (Johansen & Mayer, 1990), and with growing global food demand (Tilman et al., 2001), alternative pest control techniques need to be employed in order to minimize environmental impact of agricultural intensification. Pollination by honey bees is currently estimated to have a value of $1 billion per year in Canada and $15 billion per year in the United States (Alberta Agriculture, Food, and Rural Development, 2005). The direct economic losses of honey bee colonies and honey production attributable to pesticide use has been estimated to be $40 million/year, but the agricultural losses due to reduction in pollination by honey bees may be around $4 billion/year (Pimentel et al., 1992). Although difficult to quantify using our present knowledge, pesticides are having a serious impact on wild bee populations as well, with significant losses to biodiversity and crop yields.

Bt toxins that are now being used in commercial GM crops are not harmful to bees, and therefore use of Bt crops could result in lower use of synthetic pesticides, thus reducing harm to bee populations. The use of Bt cotton varieties have reduced synthetic pesticide applications by one-third (Qaim, 2003) to one-half (Qaim & De Janvry, 2005) in India and Argentina. Because cotton farming is the single largest user of insecticides worldwide, widespread adoption of Bt cotton could help reduce global use of synthetic pesticides. Another potential benefit of GM insecticidal crops is that the pesticide is produced by the plant and will not drift to surrounding areas.

Indirect Negative Effects of GM Crops

Transformation Effects

Introduction of transgenes to plants could result in unintended phenotypic changes that cause the GM line to be either less attractive or less nutritious to bees than the equivalent non-GM lines. Two types of indirect changes in plant phenotype due to transformation are outlined by Malone and Pham-Delegue (2001). First, an insertional mutagenesis event could occur during transformation, causing changes such as decreased nectar or altered pollen composition. If the change is noted, GM lines could be selected that do not have altered traits. Second, pleiotropic effects could occur, in which the transgene product interferes with a biochemical pathway in the plant, resulting in a phenotypic change. If the change is not noticeable, line selection cannot be used to prevent the phenotypic change resulting from transformation.

Little is known about such subtle alterations, and we lack information for most transgenic lines on whether they are different in ways that could have an impact on bee populations. One study of honey bees and transgenic oilseed rape showed a difference in nectar in the transformed line but no difference in bee foraging behavior on conventional and genetically modified plants (Picard-Nizou et al., 1995). Yet, to determine whether line effects are present, each transformed line, prior to commercialization, should be analyzed for changes in nectar and pollen composition. If so, field studies of bee activity in GM versus conventional varieties could indicate further whether the GM line was in some way negatively affecting bees. Such testing will be most important for bee-pollinated crops.

Agroecosystem Effects

In 2005, approximately 71% of the global GM acreage was planted with herbicide-tolerant crops, with an additional 11% planted with crops having both herbicide-tolerance and pest-resistance transgenes (James, 2005). Herbicide-tolerant crops are marketed to growers as a technology to reduce weed abundance in fields resulting in increased yields. Effective weed control in many conventional, non-GM crops is difficult because once the crop has emerged, growers are limited to herbicides that will not damage the crop. Crop varieties that are resistant to herbicides are very attractive to growers because they make it possible for them to control weeds even after crop emergence.

It is not likely that herbicide-tolerant crops will cause direct harm to managed or wild bees (Huang et al., 2004), but within treated areas changes in weed abundance and diversity could decrease food resources for wild bees. Most of the concern regarding consequences of GM-herbicide tolerant crops on wild bees stems from evidence suggesting that bees are more abundant and diverse in areas with greater plant diversity and abundance, but some recent studies indicate that changes in agricultural practices associated with GM herbicide-tolerant crops do indeed have significant negative impacts on bee populations through such ecosystem-level effects. Farm-scale evaluations (FSE) were conducted over a 3-year period in the United Kingdom for the purpose of evaluating effects to wildlife resulting from the way GM herbicide-tolerant crops are managed in comparison with conventional, non-GM varieties (Squire et al., 2003). Weed biomass was lower in GM herbicide-tolerant beet and rape (Hawes et al., 2003), and bee densities were lower in these GM herbicide-tolerant crops than in conventional fields (Haughton et al., 2003). Lower bee abundance in the GM herbicide-tolerant fields does not necessarily mean smaller populations in the region but could reflect foraging choices that bees are making (Hawes et al., 2003).

Smaller bee populations in GM herbicide-tolerant crops could result not only in long-term aggravation of current declines of wild bee populations in agroecosystems but also in poorer crop yields and economic returns. Crops such as canola (*Brassica* spp.) that benefit from bee pollination have lower bee populations and poorer seed set in GM herbicide-tolerant fields when compared to conventional and organic varieties (Morandin & Winston, 2005; figure 12.1). Morandin and Winston (2005) found that mean percent seed set in GM herbicide-tolerant fields was only 78% of the total potential seed set, a loss of about six seeds per pod due to lack of pollen transfer, whereas organic fields had seed set rates of 99%. Correspondingly, bee abundances

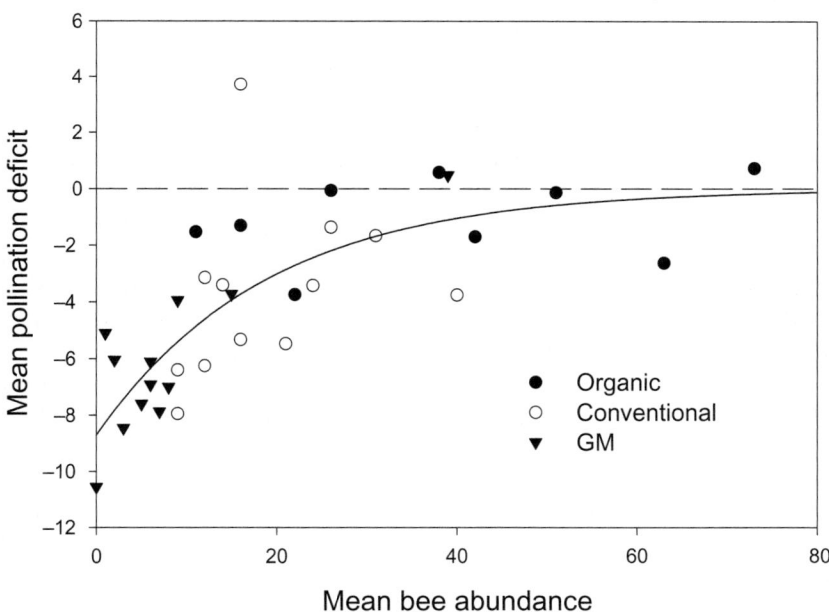

Figure 12.1 A possible indirect agroecosystem effect of genetically modified herbicide-tolerant canola on bees and crop pollination. Pollination deficit is a measure of the difference between actual seed set and potential seed set with full pollination. Note that there are fewer bees in the genetically modified fields, corresponding to lower actual seed set (from Morandin & Winston, 2005).

were approximately three times higher in organic fields than in GM herbicide-tolerant canola fields. Conventional fields had bee numbers and seed set that were intermediate to organic and GM herbicide-tolerant canola fields. Organic fields differ from conventional and GM herbicide-tolerant fields in many respects including field size, pesticide use, and species of canola. However, conventional fields were of similar size to GM herbicide-tolerant fields, had the same species of canola, and were farmed in a similar manner. Because of this and a positive correlation between weed cover and bee abundance, we hypothesized that the pattern of increasing bee abundance from GM herbicide-tolerant to conventional to organic fields was at least partially a result of increasing weed cover. Further interpretation of these data has indicated that both surrounding land and weed cover play a role in determining bee abundance in fields (Morandin & Winston, unpublished data).

Although some evidence suggests that GM herbicide-tolerant crops are associated with lower wild bee populations and poorer relative seed set, strategies to manage this potential economic and environmental problem are possible. Fewer weeds in fields are one of the potential benefits of GM herbicide-tolerant crop varieties because lower weed abundance in fields increases crop yield (Howatt, 2006). However, if extremely low weed cover causes low bee abundance in fields, bee-pollinated GM herbicide-tolerant crops may not result in any yield benefit over more "weedy" cropping methods. Rather than adopting methods that are more likely to result in higher densities of weeds in fields, a strategy that is unlikely to be popular with many growers, it may be possible for larger bee abundances to be encouraged in low-weed fields through management of surrounding land.

Morandin and Winston (2006), using data collected over a 2-year period in conventional and GM herbicide-tolerant canola fields, suggested that wild bee populations could be increased by leaving some land around fields uncultivated. We developed a model of canola profit in a 576 ha area, based on the percentage of land devoted to canola and uncultivated land. The model simulated profit along a continuum, from canola fields covering the entire area to all 576 ha being uncultivated land. Profit from canola increased when the percentage of uncultivated land was increased from 0 to 30%. When all of the area was devoted to canola and no area left uncultivated, bee abundance was low in the model, resulting in relatively low seed set. As uncultivated land increased from 0 to 30%, bee abundance increased, resulting in increasing seed set, yield, and profit. As uncultivated land was increased above 30%, profit from canola began to decrease because benefits from greater pollination were outweighed by loss of cultivated land. Therefore, we found an uncultivated land economic optimum of 30% in our study area. The model simulated a closed system that was either canola or uncultivated land and did not extend to multiyear situations or profit if other crops were present. Despite these constraints, the model reveals that some natural land in agroecosystems could make economic sense, as well as aiding biodiversity. The model has been adapted by M. Wonneck (Prairie Farm Rehabilitation Administration, Agriculture and Agri-food Canada) to incorporate a 4-year crop rotation with 3 years of nonpollinator-dependent crops (Wonneck, personal communication, February 2007). He found that landscape profit was maximized when 15% of the land was uncultivated.

Altering weed abundance in and around fields likely has varying effects on different wild bee species. Although currently we have no data on different species of wild bees and indirect agroecosystem effects of GM herbicide-tolerant crops, some speculation is possible. Larger bees, such as bumble bees, tend to forage greater distances than other bees and are able to forage on a wide range of flower types. Therefore, reduction of long-term forage in GM herbicide-tolerant fields may not affect bumble bee populations as much as those of smaller bees that are unable to forage longer distances to find food.

Maintenance or restoration of natural land in agricultural landscapes preserves wild bee populations and pollination; however, in many cases, it may not be practical or achievable. Despite knowledge and government support of natural areas in agroecosystems, the introduction and maintenance of natural land has not been widely embraced by landowners in the United States (Lovell & Sullivan, 2006). Local high-intensity management, characteristic of GM herbicide-tolerant agriculture, could, however, be compensated for by having structurally "complex" landscapes, incorporating mosaics of agricultural types (Tscharntke et al., 2005). We assessed abundance and diversity of wild bees in GM herbicide-tolerant canola fields within structurally complex landscapes, composed of pastureland and tilled crops, and within "simple" landscapes, composed of primarily tilled crops (Morandin et al., 2007). Wild bee abundance was greater, and we found a trend toward greater bee diversity in fields within complex landscapes than in simple landscapes. Thus bee conservation can be achieved in the face of locally intense agriculture if landscapes are complex.

Indirect Effects That Are Beneficial

Many wild bees nest in the ground, and tilling of fields could disrupt these nest sites. Tillage of annual crops often is done either in the fall after the crop has been harvested or in the spring before the new crop is seeded. The purpose is to loosen the soil in order to make planting easier and to disrupt weed growth. Growers can spray broad-spectrum herbicides after crop emergence with GM herbicide-tolerant crops such as canola so they do not need tillage to control weeds. Potentially this could result in more nesting habitat in fields and field margins. No studies have reported the effects of reduced tillage on wild bee populations, but in one year of our studies, four untilled fields with GM herbicide-tolerant canola did not have greater bee abundance than the tilled GM herbicide-tolerant fields. In addition, many growers are not yet taking advantage of the no-till system in GM herbicide-tolerant fields because direct seeding of a field that has not been tilled requires larger, more expensive machinery than regular seeding.

Risk Assessment of GM Crops to Bees

It is essential that all new genetically modified crop lines be tested for safety to bees, prior to commercial use. But it is important that the risks of GM crops be assessed relative to feasible alternatives. For example, a transgenic pesticidal protein that proves to be harmful to bees could be less harmful than an alternative synthetic pesticide that is being used on the non-GM crop. In addition, agroecosystem-level impacts of GM crops should be compared with effects that other agroecosystems (conventional and organic) have on bees (e.g., FSE; Champion et al., 2003; Morandin & Winston, 2005).

New lines that incorporate transgenes that are likely to be safe to bees or that have been tested and deemed safe should be checked for line effects such as pollen or nectar alteration (Pierre et al., 2003; figure 12.2). If no line effects are evident, then we should question whether ecosystem-level impacts from farming practices associated with the GM crop could negatively affect bee communities or pollination. Generally, large-scale field studies need to be performed in order to accurately measure ecosystem-level impacts on bee communities. If no negative effects are found in any of these steps, then there likely will not be impacts on bee communities greater than what can be expected with disruption from conventional agriculture. If the transgene causes lethal or sublethal impacts on bees in laboratory or field experiments, steps should be taken to minimize these impacts, such as developing lines that have lower or no transgene expression in pollen or using alternative transgenes that do not cause toxic effects. If agroecosystem effects are found, such as the lower bee abundance and pollination in some GM herbicide-tolerant crops, management of surrounding habitat could help in maintaining bee abundance and diversity.

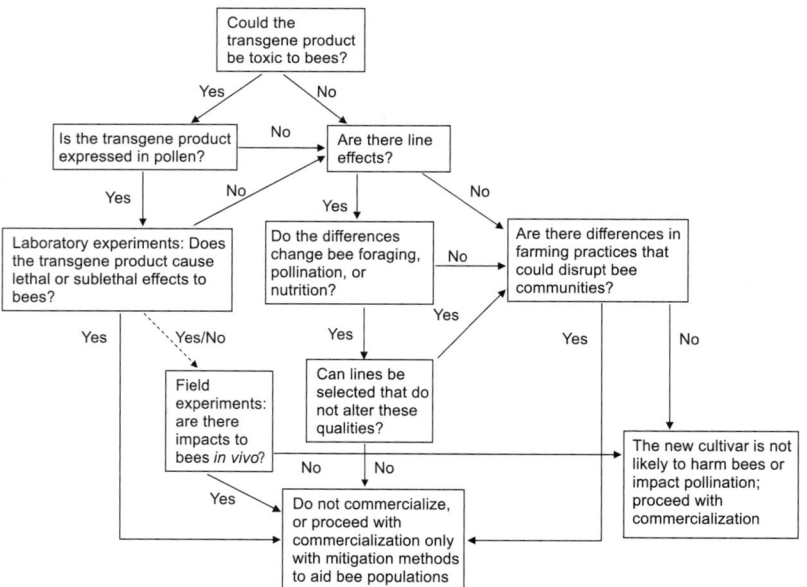

Figure 12.2 Risk assessment flow diagram for commercial production of genetically modified crops.

Conclusion

The world is demanding increased food production, and the agricultural industry is attempting to increase yields by crop improvement with genetic modification. GM crop use has increased each year since the mid-1990s, and agricultural forecasters do not expect this trend to subside for many more years. Current GM insect-resistant crops are less toxic to wildlife than many conventional pesticides and, in some cases, reduce pesticide use, potentially leading to healthier bee populations and better pollination. But these benefits are dependent on the transgene product being relatively nontoxic to bees or not expressed in pollen. If future commercialized products are toxic to bees, then their benefit will be dependent on their relative impact with respect to alternative synthetic pesticides. GM herbicide-tolerant crops are less likely to have direct impacts on bees but could negatively affect bee populations and crop pollination indirectly. GM herbicide-tolerant crops are marketed to growers for the purpose of increasing yield output through weed control, yet such weed control may negatively affect wild bee populations, and if the crops need to be insect pollinated, lower seed set could negate benefits of the herbicide-tolerant technology. If research can identify some of these potential problems and solutions, agriculturalists could take appropriate agroecosystem-level management actions, resulting in gains to both yield and biodiversity.

Acknowledgments

I thank Mark Winston and Patrick O'Hara for comments on an earlier draft of this chapter. Funding for my research was provided by the British Columbia Honey Producers Association, a research partnership agreement with the Natural Sciences and Engineering Research Council of Canada (NSERC), and associated industry collaborators: Bayer CropScience, Monsanto, Dow AgroSciences Canada, Inc., and Crompton Co./Cie, an NSERC postgraduate scholarship to L. Morandin and an NSERC Discovery Grant to Mark Winston.

References

Alberta Agriculture, Food, and Rural Development. (2005, May 23). Serious threat to Canadian honey bees. *Agri-News*. Retrieved June 2007, from http://www1.agric.gov.ab.ca/$department/newslett.nsf/pdf/agnw7331/$file/May%2023.pdf.

Benz, G., & Joeressen, H. J. (1994). A new pathotype of *Bacillus thuringiensis* with pathogenic action against sawflies (Hymenoptera, Symphyta) (Bulletin OILB-SROP 17). Montfavet, France: OILB.

Beversdorf, W. D. (1993). *Traditional crop breeding practices: An historical review to serve as a baseline for assessing the role of modern biotechnology*. Paris: Organization for Economic Co-operation and Development.

Brodsgaard, H. F., Brodsgaard, C. J., Hansen, H., & Lovei, G. L. (2003). Environmental risk assessment of transgene products using honey bee (*Apis mellifera*) larvae. *Apidologie, 34*, 139–145.

Cartagena Protocol on Biosafety to the Convention on Biological Diversity. (2000). Retrieved June 2007, from http://www.biodiv.org/doc/legal/cartagena-protocol-en.pdf.

Champion, G. T., May, M. J., Bennett, S., Brooks, D. R., Clark, S. J., Daniels, R. E., et al. (2003). Crop management and agronomic context of the farm scale evaluations of genetically modified herbicide-tolerant crops. *Philosophical Transactions of the Royal Society of London: Series B. Biological Sciences, 358*, 1801–1818.

Chrispeels, M. J., & Sadava, D. E. (2003). *Plants, genes, and crop biotechnology* (2nd ed.). Sudbury, MA: Jones & Bartlett.

Dechaume-Moncharmont, F. X., Azzouz, H., Pons, O., & Pham-Delegue, M. H. (2005). Soybean proteinase inhibitor and the foraging strategy of free flying honeybees. *Apidologie, 36*, 421–430.

Gill, S. S., Cowles, E. A., & Pietrantonio, P. V. (1992). The mode of action of *Bacillus thuringiensis* endotoxins. *Annual Review of Entomology, 37*, 615–636.

Hanley, A. V., Huang, Z. Y., & Pett, W. L. (2003). Effects of dietary transgenic Bt corn pollen on larvae of *Apis mellifera* and *Galleria mellonella*. *Journal of Apicultural Research, 42*, 77–81.

Haughton, A. J., Champion, G. T., Hawes, C., Heard, M. S., Brooks, D. R., Bohan, D. A., et al. (2003). Invertebrate responses to the management of genetically modified herbicide-tolerant and conventional spring crops: II. Within-field epigeal and aerial arthropods. *Philosophical Transactions of the Royal Society of London: Series B. Biological Sciences, 358*, 1863–1877.

Hawes, C., Haughton, A. J., Osborne, J. L., Roy, D. B., Clark, S. J., Perry, J. N., et al. (2003). Responses of plants and invertebrate trophic groups to contrasting herbicide regimes in the farm scale evaluations of genetically modified herbicide-tolerant crops. *Philosophical Transactions of the Royal Society of London: Series B. Biological Sciences, 358:* 1899–1913.

Hofmann, C., & Luthy, P. (1986). Binding and activity of *Bacillus thuringiensis* delta-endotoxin to invertebrate cells. *Archives of Microbiology, 146*, 7–11.

Howatt, K. A., Endres, G. J., Hendrickson, P. E., Aberle, E. Z., Lukach, J. R., Jenks, B. M., et al. (2006). Evaluation of glyphosate-resistant hard red spring wheat (*Triticum aestivum*). *Weed Technology, 20*, 706–716.

Huang, Z. Y., Hanley, A. V., Pett, W. L., Langenberger, M., & Duan, J. J., (2004). Field and semifield evaluation of impacts of transgenic canola pollen on survival and development of worker honey bees. *Journal of Economic Entomology, 97*, 1517–1523.

Ingram, M., Nabhan, G. P., & Buchmann, S. (1996). Our forgotten pollinators: Protecting the birds and bees. *Global Pesticide Campaigner, 6*. Retrieved March 2006, from http://www.pmac.net/birdbee.htm.

James, C. (2005). *Global status of commercialized biotech/GM crops* (ISAAA Brief No. 34). Ithaca, NY: International Service for the Acquisition of Agri-Biotech Applications.

Johansen, C. A., & Mayer, D. F. (1990). *Pollinator protection: A bee and pesticide handbook*. Cheshire,CT: Wicwas Press.

Liu, B., Xu, C. G., Yan, F. M., & Gong, R. Z. (2005). The impacts of the pollen of insect-resistant transgenic cotton on honeybees. *Biodiversity and Conservation, 14*, 3487–3496.

Losey, J. E., Rayor, L. S., & Carter, M. E. (1999). Transgenic pollen harms monarch larvae. *Nature, 399*, 214–214.

Lovell, S. T., & Sullivan, W. C. (2006). Environmental benefits of conservation buffers in the United States: Evidence, promise, and open questions. *Agriculture, Ecosystems and Environment, 112,* 249–260.

Malone, L. A., & Pham-Delegue, M. H. (2001). Effects of transgene products on honey bees (*Apis mellifera*) and bumblebees (*Bombus* sp.). *Apidologie, 32,* 287–304.

Malone, L. A., Todd, J. H., Burgess, E. P. J., & Christeller, J. T. (2004). Development of hypopharyngeal glands in adult honey bees fed with a Bt toxin, a biotin-binding protein and a protease inhibitor. *Apidologie, 35,* 655–664.

Matson, P. A., Parton, W. J., Power, A. G., & Swift, M. J. (1997). Agricultural intensification and ecosystem properties. *Science, 277,* 504–509.

Michener, C. D., McGinley, R. J., & Danforth, B. N. (1994). *The bee genera of North and Central America (Hymentoptera: Apodia).* Washington, DC: Smithsonian Institution Press.

Morandin, L. A., & Winston, M. L. (2003). Effects of novel pesticides on bumble bee (Hymenoptera: Apidae) colony health and foraging ability. *Environmental Entomology, 32,* 555–563.

———. (2005). Wild bee abundance and seed production in conventional, organic, and genetically modified canola. *Ecological Applications, 15,* 871–881.

———. (2006). Pollinators provide economic incentive to preserve natural land in agroecosystems. *Agriculture Ecosystems and Environment, 116,* 289–292.

Morandin, L. A., Winston, M. L., Abbott, V. A., & Franklin, M. T. (2007). Can pastureland increase wild bee abundance in agriculturally intense areas? *Basic and Applied Ecology, 8,* 117–124.

Paoletti, M. G., & Pimentel, D. (2000). Environmental risks of pesticides versus genetic engineering for agricultural pest control. *Journal of Agricultural and Environmental Ethics, 12,* 279–303.

Picard-Nizou, A. L., Grison, R., Olsen, L., Pioche, C., Arnold, G., & Pham-Delegue, M. H. (1997). Impact of proteins used in plant genetic engineering: Toxicity and behavioral study in the honeybee. *Journal of Economic Entomology, 90,* 1710–1716.

Picard-Nizou, A. L., Pham-Delegue, M. H., Kerguelen, V., Douault, P., Marilleau, R., Olsen, L., et al. (1995). Foraging behaviour of honey bees (*Apis mellifera* L.) on transgenic oilseed rape (*Brassica napus* L. var. *oleifera*). *Transgenic Research, 4,* 270–276.

Pierre, J., Marsault, D., Genecque, E., Renard, M., Champolivier, J., & Pham-Delegue, M. H. (2003). Effects of herbicide-tolerant transgenic oilseed rape genotypes on honey bees and other pollinating insects under field conditions. *Entomologia Experimentalis et Applicata, 108,* 159–168.

Pimentel, D., Acquay, H., Biltonen, M., Rice, P., Silva, M., Nelson, J., et al. (1992). Environmental and economic costs of pesticide use. *BioScience, 42,* 750–760.

Prescott-Allen, R., & Prescott-Allen, C. (1990). How many plants feed the world? *Conservation Biology, 4,* 365–374.

Qaim, M. (2003). Bt cotton in India: field trial results and economic projections. *World Development, 31,* 2115–2127.

Qaim, M., & De Janvry, A. (2005). Bt cotton and pesticide use in Argentina: Economic and environmental effects. *Environment and Development Economics, 10,* 179–200.

Sagili, R. R., Pankiw, T., & Zhu-Salzman, K. (2005). Effects of soybean trypsin inhibitor on hypopharyngeal gland protein content, total midgut protease activity and survival of the honey bee (*Apis mellifera* L.). *Journal of Insect Physiology, 51:* 953–957.

Sears, M. K., Hellmich, R. L., Stanley-Horn, D. E., Oberhauser, K. S., Pleasants, J. M., Mattila, H. R., et al. (2001). Impact of Bt corn pollen on monarch butterfly populations: A risk assessment. *Proceedings of the National Academy of Sciences of the USA, 98*, 11937–11942.

Simpson, R. M., Burgess, E. P. J., & Markwick, N. P. (1997). *Bacillus thuringiensis* delta-endotoxin binding sites in two lepidoptera, *Wiseana* spp. and *Epiphyas postvittana*. *Journal of Invertebrate Pathology, 70*, 136–142.

Squire, G. R., Brooks, D. R., Bohan, D. A., Champion, G. T., Daniels, R. E., Haughton, A. J., et al. (2003). On the rationale and interpretation of the farm scale evaluations of genetically modified herbicide-tolerant crops. *Philosophical Transactions of the Royal Society of London: Series B. Biological Sciences, 358*, 1779–1799.

Tilman, D., Fargione, J., Wolff, B., D'Antonio, C., Dobson, A., Howarth, R., et al. (2001). Forecasting agriculturally driven global environmental change. *Science, 292*, 281–284.

Tscharntke, T. A., Klein, M., Kruess, A., Steffan-Dewenter, I., & Thies, C. (2005). Landscape perspectives on agricultural intensification and biodiversity: Ecosystem service management. *Ecology Letters, 8*, 857–874.

U.S. Census Bureau. (2008.) World population information. Retrieved February 2008, from http://www.census.gov/ipc/www/idb/worldpopinfo.html.

U.S. Environmental Protection Agency. (2000). October 18–20, 2000, FIFRA SAP meeting: Bt plant pesticides risk and benefits assessment. Retrieved June 2007, from http://www.epa.gov/oscpmont/sap/meetings/2000/october/questions.pdf.

———. (2001). Biopesticides registration action document: *Bacillus thuringiensis* plant-incorporated protectants. Retrieved June 2007, from http://www.epa.gov/pesticides/biopesticides/pips/bt_brad.htm.

Wilkinson, J. E., Twell, D., & Lindsey, K. (1997). Activities of CaMV 35S and nos promoters in pollen: Implications for field release of transgenic plants. *Journal of Experimental Botany, 48*, 265–275.

13 The Future of Agricultural Pollination

Rosalind R. James and Theresa L. Pitts-Singer

We hope that we have demonstrated to you the many ways that bees affect agriculture. Most notable, of course, is pollination. Worldwide, approximately 20% of all food-crop production and about 15% of seed crops require the help of pollinators for full pollination (Klein et al., 2007). These food crops include delicious products such as many of our tree fruits (e.g., apples, cherries, mangos, and avocados) and nuts (e.g., almonds, peanuts, macadamia nuts, and Brazil nuts), plus squashes, cucumbers, melons, citrus, and berries. In addition, many of our oil seed crops benefit from bee pollination, such as canola, safflower, coconut, and so forth.

The honey bee, *Apis mellifera*, is the most commonly used agricultural pollinator in North America and Europe and in many other parts of the world, as well. Farmers in the United States rent more than 2 million honey bee colonies every year for pollination, but the honey bee is increasingly being threatened by a myriad of problems. Varroa mites, pesticide resistance in these mites, viral diseases transmitted by varroa mites, antibiotic resistance in American foulbrood, tracheal mites, and small hive beetles are critical problems in honey beekeeping that have accumulated to a point at which it is not as economical to manage this bee as it once was. For example, the demand for honey bees in California almond orchards is rapidly increasing due to increasing acreage, but this is occurring during a time when honey bee colonies are not easy to produce, and, as a result, the cost of renting bees is up from US$50 per hive in 2003 to US$140 per hive in 2006 (Sumner & Boriss, 2006). That is a nearly threefold price increase in 4 years for a crop that requires five to seven hives per hectare.

In addition, not all crops are well pollinated by honey bees. For example, tomatoes require buzz pollination, which honey bees cannot achieve; alfalfa flowers are not properly worked by honey bees, and honey bees do not work well in greenhouses and under row covers. Fortunately, honey bees are not the only bees that make good pollinators. Approximately 16,000 species of bees are known in the world (Michener, 2000); only a few species, however, are managed or "kept" specifically as crop pollinators. These include some bumble bees (*Bombus terrestris* and *B. impatiens*), the alfalfa leafcutting bee (*Megachile rotundata*), a few species of mason bees (*Osmia* spp.), and the alkali bee (*Nomia melanderi*). The potential exists to manage many more species of bees and to broaden the use of those that are already being used. However, the management of non-*Apis* pollinators poses its own set of problems. Nesting materials, shelters, and rearing techniques have been well developed for the alfalfa leafcutting bee and are commercially available, but methods for propagating other important pollinators, such as blue orchard bees and bumble bees, are less well developed. Furthermore, these bees have their own set of disease and parasite problems that need to be addressed.

Amazingly, the huge diversity of wild bees essentially has been unexplored, particularly with regard to the plants that they visit and crops that they could pollinate. The identification and classification of many important pollinator groups is inadequate, with many species still undescribed, and the relationships between species and between groups of species remains unresolved. In addition, we lack an understanding of native bee biology and of the natural enemy and pathogen complexes of these bees. Some of our native bees are threatened by a shrinking natural habitat, and a lack of information about their biology, bionomics, and natural population regulation inhibits our ability to rely on their populations for pollination. More information is needed to help protect the native bees and to open up the potential for the domestication of some of them before this treasure of diversity is lost forever.

The year 2006 was an important year for bee research worldwide. The complete genome of the honey bee was sequenced and reported in *Nature* (Honey bee Genome Sequencing Consortium, 2006), and the genomes of two major honey bee pathogens were sequenced, *Ascosphaera apis* and *Paenibacillus larvae* (Qin et al., 2006). At about the same time, a new phylogenetic analysis that included gene sequence data was used to reevaluate our previous conceptions of the evolution of bees (Danforth et al., 2006). Earlier in the year, a *Science* report documented what appears to be a major decline in bees in England and the Netherlands, possibly a 30% loss in species richness since 1980, especially among specialist bees, and a corollary decline in wild plant species that require insect pollination (Biesmeijer et al., 2006). The U.S. National Academy of Sciences (NAS) also assembled a committee to review the status of pollinators in North America, and this committee completed their report in 2006. Of course, this is only a fraction of the bee research activities reported in 2006; however, to have so many reports on bees in key scientific journals that cater to a broad audience is not common in our field. It highlights the fact that bees, and especially bees as pollinators, are now well recognized as being very important for both wild ecosystems and agricultural systems, and the public is generating a growing concern that we may lose this valuable resource.

The NAS report, *Status of Pollinators in North America* (National Research Council, 2007), highlights areas of concern that directly relate to providing adequate pollinators for agriculture and the environment, including the need to: (1) better enforce regulations designed to limit the accidental introduction of honey bee pests and pathogens and increase research on innovative ways to control honey bee pests; (2) expand research efforts to improve our ability to identify the great diversity of bee species that occupy our farmlands and wildlands so that we can further study their abundance and natural history; (3) take measures to prevent the spread of pathogens from managed to wild bee populations; (4) establish discovery surveys for pollinators of rare and endangered plants and document long-term population trends in wild bees; (5) identify the effects of habitat loss and fragmentation on bee populations; and (6) expand economic incentives for pollinator conservation while enhancing pollinator awareness through public education and professional societies.

Many of the recommendations in the NAS report focused on wild pollinators. Besides providing an ecosystem function, wild pollinators directly provide pollination services to agriculture, as discussed in several chapters in this book; they are also a source of potential new pollinators for the future. Agriculture is destined to change over time, and so will its pollination needs. In these ways, conservation of wild pollinators and agriculture cross paths. Another way that these paths intersect is that, when wild bees are adequate to meet the pollination needs of agriculturalists, we eliminate some of the social and economic pressures to introduce bees from foreign regions. Bringing new bee species into an area increases risks of accidentally introducing new bee diseases and parasites or of accidentally releasing bees that may competitively displace native bees or further compound honey bee pest management.

Before World War II, the manner of public discussions about farming and agriculture was often from the point of view of "agrarianism," but since that time, our view of agriculture has changed. Advocates of agrarianism viewed farming as a way of life that was "rewarding, a public good, and a source of moral virtue, [while] current writers on farmland preservation speak of farming almost entirely in utilitarian terms, describing its productive capacity and its economic returns" (Mariola, 2005). Mariola (2005) goes on to explain that proponents who use this "economic utilitarianism" argument for farmland preservation evaluate land use decisions based mainly on food production potential and economic criteria and, despite their good intentions, disregard the agrarian ethic, or social value, associated with farming. Wildlife conservation efforts sometimes utilize a similar economically oriented philosophy. To gain the attention of the general public and business corporations, conservationists' arguments are often made in the context of the economic value of wildlife to society. Mariola's (2005) argument for a return to agrarianism as a basis for farm preservation is reminiscent of the wildlife conservation ideal proposed by Aldo Leopold as far back as 1949, in what he called the land ethic: "A land ethic then, reflects the existence of an ecological conscience, and this in turn reflects a conviction of individual responsibility for the health of the land. Health is the capacity of the land for self-renewal. Conservation is our effort to understand and preserve this capacity" (Leopold 1948, p. 221).

Our need to preserve pollinators easily goes across the boundaries between agrarianism and a land ethic, despite the fact that the values and goals of farmers and wildlife conservationists are not always in synchrony with each other. For example, many species of bees are rare or specialists and have no known utilitarian value to agriculture. Conservation efforts generally target the preservation of rare or endangered species, and successful conservation often calls for the sacrifice of some human activity, and these activities can include agricultural practices such as pesticide use, tillage, or other land uses. Thus, if we only take economics into consideration, we will find ourselves in a position in which we sometimes must weigh the economic value of a rare or specialist bee against the economic gains of an agricultural practice that would be interrupted in order to preserve that bee. This approach to conservation ignores the value of bees to human society as a whole. It is true that bees produce a direct benefit to humans by pollinating agricultural crops and gardens, but they are also important for maintaining a balanced ecosystem, on which we are dependent and from which we can learn social values. These small, unobtrusive creatures are symbolic of our interconnection with nature and highlight the ethical responsibility we have toward maintaining the integrity of the natural world.

References

Biesmeijer, J. C., Roberts, M., Reemer, M., Ohlemüller, R., Edwards, M., et al. (2006). Parallel declines in pollinators and insect-pollinated plants in Britain and the Netherlands. *Science, 313*, 351–354.

Danforth, B. N. S., Sipes, S., Fang, J., & Brady, S. G. (2006). The history of early bee diversification based on five genes plus morphology. *Proceedings of the National Academy of Sciences of the USA, 103*, 15118–15123.

Honeybee Genome Sequencing Consortium. (2006). Insights into social insects from the genome of the honeybee *Apis mellifera*. *Nature, 443*, 931–949.

Klein, A.-M., Vaissière, B. E., Cane, J. H., & Steffan-Dewenter, I. (2007). Importance of pollinators in changing landscapes for world crops. *Proceedings of the Royal Society of London: Series B, 274*, 303–313.

Leopold, A. (1949). *A Sand County almanac and sketches here and there*. New York: Oxford University Press.

Mariola, M. J. (2005). Losing ground: Farmland preservation, economic utilitarianism, and the erosion of the agrarian ideal. *Agriculture and Human Values, 22*, 209–223.

Michener, C. D. (2000). *The bees of the world*. Baltimore: John Hopkins University Press.

National Research Council. (2007). *Status of pollinators in North America*. Washington, DC: National Academies Press.

Qin, X., Evans, J. D., Aronstein, K. A., Murray, K. D., & Weinstock, G. M. (2006). Genome sequences of the honey bee pathogens *Paenibacillus larvae* and *Ascosphaera apis*. *Insect Molecular Biology, 15*, 715–718.

Sumner, D. A., & Boriss, H. (2006). Bee-conomics and the leap in pollination fees. *Giannini Foundation of Agricultural Economics, 9*, 9–11.

INDEX

Note: numbers followed by "t" refer to tables; numbers followed by "f" refer to figures.

agrarianism, 221
agriculture, impact on wild bees, 15–21, 58, 212–214, 221–222
alfalfa,
   flower tripping mechanism, 106–107
   history, 105–107
alfalfa leafcutting bee. *See* bee species, alfalfa leafcutting
alfalfa seed, 106–107 (*see also* pollination of)
alkali bee. *See* bee species, alkali
almond. *See* pollination of
Anthophora plumipes. See bee species
antipest plastic. *See* greenhouse
Apiaceae, 50t
Apicystis bombi, 127t
Apis species. *See* bee species, honey
Apoidea/Apiformes, 5, 145, 204
apple. *See* pollination of
Ascosphaera species
  aggregata, 116, 126t, 129–130, 133
  apis, 125–126t, 133, 220
    genome, 220

  proliperda, 129–130
  torchioi, 126t
Aspergillus flavus, 126t
Asteraceae, 50t, 68
Astragalus filipes, 50t, 52, 53f, 55, 58

Bacillus species
  subtilis, 68, 73, 75
  thuringiensis, 68–69, 71, 74–75, 206–208
bacteria (*see also Bacillus* species)
  and biocontrol, 65, 68, 207
  as pathogens of bees, 125, 126t, 128, 131
Baker's Law, 168
Balsamorhiza sagittata, 50t, 53f, 60
banded sunflower moth, 67–68
Banksia ornata, 148, 150
bat, 11, 150
Beauveria bassiana
  biocontrol of tarnished plant bugs, 69f, 70f, 69–71, 75
  formulation for pollinator biocontrol vector technology, 72
  pathogenicity to bees, 71–72, 127

bee
  behavior, 30, 33, 178
    aggressive, 6, 156
    cooperative, 6
    foraging, 5, 7, 12, 18–19, 30, 36, 148–149, 151, 158, 187–188, 204–205
    in greenhouse, 30–35
    hive, 6, 128
    hygienic, 128
    -al interactions, 12, 18, 156
      pollinator, 5, 11–14, 18, 35, 60, 152, 166, 176, 178
    learning, 31, 151, 205–206
    nectar-robbing, 152
    nesting, 87–88, 114t, 156
    and pathogenicity, 128, 132f, 132
    pollen-removing, 13
    recruitment, 14, 35
    sexual, 87
    species-specific,13
  board, 8, 110–112
  cavity-nesting, 6, 8, 21, 57–58, 60–61, 156
  conservation, 213, 220–222
  definition of, 4–5
  density
    in canola, 186–188
    and plant interactions, 167, 173–174, 190, 197–198
    sampling, 115
    stocking, 60
  diseases, 124–137 (*see also* disease, of bees)
  diversity, 5, 204, 213
    agricultural impacts, 15–21, 209, 212–213
    declines, 220–221
    genetic, 128
    impact on pollination service, 18
  economic value of, 14–15, 21, 145, 146t, 204, 209, 219, 222
  ecophysiology, 83–98
  evolution, 4–5, 118–119, 220
  foraging range, 15–16, 18, 20, 114t, 175, 187, 204
  ground-nesting, 8, 20, 56–59, 61, 83, 107–110, 213

  management for pollination, 7–8, 19–21, 83–85, 94–96, 107–110, 113–117
  as a pollinator, 7, 174–176, 186–193
  preferences
    flower, 5, 54–56, 151, 155, 174
    forage, 114t
    color, 151
  sociality, 6, 14, 118, 128, 204
  solitary 6, 107
  and wildland seed propagation, 48–62
bee species
  alfalfa gray-haired, 21, 83
  alfalfa leafcutting
    vs. alkali, 114t
    chalkbrood, 116 (*see also Ascosphaera*)
    life cycle, 96–98, 108, 110–113
    management of, 110–119, 114t
    management problems, 115–117
    management using loose cells, 111–113
    multivoltinism in, 111, 116
    nest structure, 108f, 110–111
    parasite and predator control, 112, 114t, 116
    release rates, 115, 117
    "second generation," 116
    sex ratio, 111
    and wildland seed, 56, 61
  alkali
    vs. alfalfa leafcutting, 114t
    artificial nesting bed construction, 109–110
    crop pollination, 56, 115
    decline in use, 113–115, 117
    disease, 125
    economic considerations, 117–118
    impact of pesticide on, 117
    introduction to New Zealand, 156
    life cycle, 107–109
    mortality, 115
    native habitat, 107
    nest aggregations, 6
    nesting habits, 107–108, 114t
    nest structure, 108t, 107–108
    population, 113, 115
    and wildland seed, 57, 61

Anthophora plumipes, 157–158
Apis mellifera. *See* bee species, honey
blue orchard, 20–21, 55, 57, 61, 85–97, 118, 126t
bumble
  and biological control of pests, 65–75, 73f,
  Bombus ephippiatus, 154
  B*ombus hortorum*, 153
  Bombus ignitus, 34–35
  Bombus impatiens, 34–35, 134, 153–154
  Bombus lucorum, 34
  Bombus occidentalis, 34–35, 149, 151
  Bombus ruderatus, 150–151, 153
  Bombus subterraneus, 153
  B*ombus terrestris*
    effect on other pollinators, 151–153
    and flower damage, 152
    range, 151
    terrestris canariensis, 152
    terrestris sassaricus, 152
    terrestris terrestris, 152
  Bombus wilmattae, 154
  colony size, 34
  diseases, 34, 126–127t, 128, 134
  drones, 41
  economics, 34, 159
  flight speed, 186–187, 196
  floral attraction, 28–30
  life history, 6, 34–35
  management, 8, 34–35, 220
  and Nosema, 151
  restriction on use, 33–34, 134, 154–155, 157, 159
  and wildland seed, 55–57, 61
carpenter, 6, 8
*Centau*rea leafcutting, 155–156, 171, 174
Eufriesea nigrohirta, 149
Habrop*oda laboriosa*. *See* bee species, southeastern blueberry
honey, 146
  Africanized, 19, 147
  and biological control of pests, 65–75
  as pollinators, 10–12
  genome, 220
  sociality, 6
  subspecies, 146–147
  horn-faced, 39, 55, 85–89, 92, 94, 96–98, 156
  mason, 6, 8, 84, 220 (*see also* bee species, *Osmia*)
Megachile
  apicalis. See bee species, Centaurea leafcutting
  centrus, 129
  rotundata. See bee species, alfalfa leafcutting
Osmia
  aglaia, 85
  attriventris, 85
  bruneri, 57–58
  caerulescens, 85
  californica, 60
  cornifrons. *See* bee species, horn-faced
  cornuta, 85–86, 88–91, 93–96
  excavata, 85
  lignaria. *See* bee species, blue orchard
  montana, 60
  ribifloris, 85
  rufa, 85
  sanrafaelae, 57, 85
  subgenus, 85
*Pepon*apis pruinosa. See bee species, squash
Pyrobombus, 154
Rhophitoides canus. See bee species, alfalfa gray-haired
southeastern blueberry, 57
squash, 57, 174
stingless, 5, 17
  Trigona carbonaria, 16
sweat 6, 8
Xenoglossa, 174
beekeeping
  migratory, 54, 134
  losses, 54, 124, 158, 219
Bettsia alvei, 129
binab, 74 *(see also Trichoderma spp.)*

biocontrol vector technology. *See* pollinator biocontrol vector technology
Bt. *See Bacillus* species, *thuringiensis*
bivoltinism, 85, 97
blueberry, 11–12, 57, 68, 89
   mummy berry disease, 68
blue orchard bee. *See* bee species
botanigard. *See Beauveria bassiana*
Botrytis cinerea. See gray mold
Brachyloma ericoides, 148
Brassica napus. See canola
bumble bee. *See* bee species, bumble

cabbage seed weevil, 68
Callistemon rugulosus, 148, 150
canola, 68, 75 (*see also* pollination of)
carbon dioxide (CO_2), 32, 125
carpenter bee. *See* bee species, carpenter
cell provisions, 7, 56, 59, 61, 86, 88, 96, 108f, 108–109, 111, 116, 130, 135, 136f, 185, 204–206
Centaurea. See star-thistle
Chalicodoma, 152
chalkbrood, 114t, 116, 125, 126t, 127, 129–133, 135–136, 136f
cheatgrass, 48
checkered flower beetle, 116
Ceutorhynchus assimilis. See cabbage seed weevils
Cleomaceae, 50t, 51
Cleome 54, 56
   lutea, 50t, 53f
   serrulata, 50t, 53f
Clonostachys rosea, 67–68, 71, 75
clover, 17, 56, 68, 73, 106, 153, 196 (*see also Dalea, Trifolium*)
Clusia arrudae, 149
Cochylis hospes. See banded sunflower moth
coevolution of bees and flowering plants, 5
Colony Collapse Disorder (CCD), 3, 124, 133
competitive exclusion, 149
conservation
   bee, 203, 213, 222
   biological diversity, 205

   habitat, 21
   managers, 168, 176
   plant, 167, 176, 178
   pollinator, 11, 52, 167, 178, 221
   wildflower, 51–52
   wildlife, 221
conservationist, 33, 42, 221
Correa reflexa, 148, 150
Crepsis acuminata, 50t, 52
Crithidia
   bombi, 127t
   mellificae, 127t
cucumber, 28t, 28, 29t, 36, 37t, 68, 219 (see also pollination of)
Cucurbita, 37t, 40, 57, 174
Cucurbitaceae, 28

Dalea, 50t, 53f, 56, 60–61
   ornatum, 50t
   purpurea, 53f, 56, 60–61
   searlsiae, 50t
death camus, 5
developmental biology, 84, 86, 94
diapause, 89–95, 97–98, 111
   summer, 94
   winter, 93, 95, 97–98
Didymella bryoniae, 68
disease, of bees 124–137
   and behavioral fever, 128
   chalkbrood 116, 125, 126t, 129–130, 132–133, 135–136 (see also chalkbrood)
   control of, 111, 114t, 115, 125, 127, 129–137, 132f, 154, 157
   flagellate, 127t
   gregarine, 127t
   life cycle of, 131–133, 137
   management of, 129, 137
   Nosema, 154
      apis, 126t
      bombi, 34, 126t, 134, 151
   infection, 130, 133–134, 154–155
   pathogenic organisms, 126–127t, 128
      as an epizootic element, 131–132, 132f
   stone brood, 126t

Streptococcus pluton, 126t
  survival of, 127, 131
  triangle, 131–133, 132f
  vectors, 118, 133
  virulence of pathogens, 127–128, 131
dispersal
  bee, 19, 94, 117
  exotic plants, 172
  gene, 194
  pollen, 148
  spore, 135
displacement, 150, 152–153, 156, 159
dysentery, 126t

early-flying bee populations, 86, 88, 90–91, 93f, 95, 98
economic utilitarianism, 221
economic value of pollination, 14–15, 21
egg, development, 5, 86, 88, 92, 97–98, 111
eggplant, 28t, 28, 32, 36, 37t
emergence
  period, 34, 86f, 86–87, 91, 82, 115–116
  second generation, 116
  timing (synchrony), 8, 57, 84, 89, 93–98, 111–114, 114t
  tray, 112
endangered plants, 52, 221
environmental impact, of exotic bees, 145–159
Epacris impressa, 151
epizootic, 124, 128–132
Erio*gonum umbellatum*, 50t, 53f
Erwinia amylovora. See fire blight
Euglossini, 7, 149
eusociality, 6, 118
extra-early horticulture, 27, 29t, 37t, 42

Fabaceae, 50t, 68, 105, 152–153, 172
fat body depletion, 91–95
fire
  suppression, 48
  wildfire, 48–49
fire blight, 68

flagellate disease. *See* disease, of bees
floral host, 54, 60, 174
  of alkali bee, 56
  populations, 60
  preferences, 38, 55
flower constancy, 13, 35–36
formaldehyde, 135
  para- 136
foulbrood, 124
  American, 126t, 133–136, 219
  European, 126t
fungi
  associated with bee nests, 116, 125, 128, 136f (see also Ascosphaera)
  and biocontrol, 65, 67–73
  as pathogens of bees, 126t, 128, 134–136 (*see also* diseases, of bees)
Frankliniella occidentalis. See thrips
fruit set, 12–13, 16–17, 36, 39–40, 52, 148, 171

gene flow, 184–199
genetic diversity, 51, 128
genetically modified (GM) crops
  definition and history, 203–204
  environmental impacts, 209
  gene flow, 193–198
  impacts on bees, 203–215
  public response, 184–185, 207
  regulation by government, 205–206
gray mold, 67–68, 71, 74
Great Basin, 48–52, 56, 60–61, 106
green bean, 28t, 28
greenhouse
  antipest plastic, 31f
  biological control in, 69–71, 70f
  crop, 27–28, 69, 71, 74–75, 85, 151–152, 154
  cultivation acreage, 27
  pollination in, 28–41
  whitefly. *See* whitefly
green peach aphid, 71

gregarine disease. *See* disease
habitat
fragmentation and bees, 16f, 16–21
loss, 52, 221
Halictidae, 156 (*see also* bee species, alkali bee and sweat bee)
Halictinae, 6
haplodiploidy, 5
Hedysarum boreale, 50t, 50, 52, 53f, 55, 58
Helicosmia, 85–86
Heliothis nuclear polyhedrosis virus (NPHV), 68, 72–75
hive rental, 21, 51, 54–55, 118, 147–148, 219
honey bees. *See* bee species, honey

Hoplitis, 57, 61
horn-faced bee. See bee species, horn-faced
host
acceptance, 116
of biocontrol agents, 72, 74, 159
for pathogen, 126–127t, 127, 131, 132f
plants, 38, 93, 156, 174
response, 127
specialization, 174
susceptibility, 132f, 132
hybridization, 51, 185

immune response, 128–129
Impatiens capensis, 149
importation regulations, 154, 157–158, 166, 205, 220
incubation, 57, 86f, 86–87, 91, 93, 95–98, 111–112, 114t, 116, 135
infection
bacterial, 131
co-, 129
chalkbrood. See disease, of bees
dis-, 136
from bee-vectored agents, 70f, 71
initiation, 131
Nosema. See disease, of bees
sensitivity, 127–129
sublethal, 131, 134

suppression of, 68
integrated pest management (IPM), 30, 41–42, 137
Intermountain Region, U.S., 49f, 50, 50t
Invasive Alien Species Act, 157
invasive
grass, 48
mutualism, 152, 156, 158–159, 166
plant, 166f, 166–178
weeds, 48, 169f, 169–172

killer yeast, 68

larval development, of bees, 86f, 88, 208
larval mortality, of bees, 116, 208
late-flying populations of bees, 86–88, 90–91, 93f, 94, 98
Lomatium
dissectum, 50t, 53f, 56
triternatum, 50t
loose cell bee management system, alfalfa leafcutting bees, 111, 113, 114t, 116, 136
Lupinus
argenteus, 50t
sericeus, 50t, 58
Lygus lineolaris. *See* tarnished plant bug

Malvaceae, 50t
Masaridae, 54
Medicago sativa. *See* alfalfa
Megachile rotundata. *See* bee species, alfalfa leafcutting
Megachilidae, 152, 155 (*see also* bee species, *Megachile* and *Osmia*)
Meligethes aeneus. *See* pollen beetle
Meliphagidae, 150
melon, 28t, 28, 29t, 32, 37t, 85, 114t, 118, 219 (*see also* pollination of)
Metarhizium anisopliae, 125
for biocontrol of cabbage weevils, 68, 75
for biocontrol of pollen beetles, 68, 75
pathogenicity to bees, 72
methyl bromide, 135

Metschnikovia fructicola. *See* killer yeast
microbial control, 66, 74–75, 127
microbial survey, 125
microsporidia, 126t, 128, 134
Monilinia vaccinii-corymbosi. *See* blueberry, mummy berry disease
Monoica apis, 127t
Myzus persicae. *See* green peach aphid

native plant communities
    rehabilitation of, 48–52, 61
nectar
    attraction to, 7, 13, 33, 40, 54
    composition, 7, 28–29, 29t, 38, 40, 206–207
    guides, 29–30
    quantities produced, 173, 196, 198
    robbery, 11, 151–152
nectarivorous
    bat, 150
    bird, 150
    insect, 54
nest
    aggregation, 6, 87
    building
        alfalfa leafcutting bee, 108f, 111, 117
        alkali bee, 107–109, 108f
    cell construction, 108, 110
    provisioning of, 56, 59, 61, 108
    removal from boards, 135
    sites, 6, 8, 16f, 16, 20–21, 58–59, 150, 156, 175, 213
        of alkali bees, 56–57, 107–110, 108f, 113–115, 117–118, 156–157
        of mason bees, 20, 83, 87–88
Nomia melanderi. *See* bee species, alkali
Nosema. *See* disease, of bees
nuclear polyhedrosis virus. *See Heliothis* nuclear polyhedrosis virus
nut production, 118

oligolecty, 5
    oligolege, 155–156, 174
    orchard (*see also* pollination of)
        bloom period, 87, 94, 98

yields, 98
organic farming
    vs. conventional, 20–21
    and genetically modified crops, 211–212

Paenibacillus larvae, 126t, 134, 136, 220
paraformaldehyde. *See* formaldehyde
parasite
    control, 114t (*see also* disease, of bees)
    ecto, 155
    floral, 148, 152
    insect, 14, 20, 35, 59, 114–115, 154, 157, 220–221
    myco, 68 (*see also Trichoderma*)
parthenocarpy, 36, 37t, 41
pathogens. *See* disease, of bees
pear, 68 (*see also* pollination of)
Penstemon speciosus, 50t, 53f, 54–55
pepper, 28t, 28, 29t, 38–39, 69–70, 70f, 71, 75 (*see also* pollination of)
pest control, 3–4, 16f, 30, 41, 65–75, 125, 127, 137, 209. (*see also* Integrated Pest Management)
pesticide/insecticide, 8, 16, 27, 35, 41, 75, 83, 113, 115, 117, 124, 204–206, 208–209, 212, 214–215, 219, 222 (*see also* safety)
    dispersed by bees, 67
    and pollinators, 12, 115, 117
phenology
    developmental, 90, 93
    flowering, 20, 57, 96, 167f, 173
    of *Osmia*, 57, 86f, 86, 94, 98
    of pollinators, 84, 167
Phylidonyris novaehollandiae, 150
Plantaginaceae, 50t
plant-pollinator driven meltdown, 174
plant-pollinator interaction, 4, 30, 148, 166–178, 186–188, 195–196, 199, 205
pollen
    composition, 7, 210
    odor, 38
pollen ball, 116

pollen beetle, 68
pollen wasp, 54
pollination
　buzz, 34, 38, 219
　cross-, 7, 38, 53f, 184–185, 190, 193–195, 197–198
　ecology, 51
　efficacy, 7, 56, 110, 196
　generalists, 18–19, 59, 147, 171, 175
　rate, 12, 155
　self-, 11, 13, 36, 38–39, 189–192, 196, 198
　　in crops, 189–190
　　and pollinators, 13
　　and weediness, 168, 170
　specialists, 54, 57, 59, 105, 174, 220–222
pollination by
　birds, 148, 150–151
　honey bees, 10
　managed bees, 8, 10, 220
　multiple pollinators, 12–13, 167, 178
　wild bees, 8–9, 10–21
pollination of
　alfalfa, 12, 107–117, 219
　　alkali bee vs. alfalfa leafcutting bee, 114
　almond, 18, 35, 57, 90, 148, 155, 219
　apple, 35, 55, 96, 98
　blueberry, 11–12, 57
　canola, 75, 118, 186–189, 192, 195–197, 204, 211f, 211–213
　Centaurea solstitialis. *See* pollination of, star-thistle, yellow
　coffee, 15–16, 19
　cranberry, 12, 187, 199
　cucumber, 36, 174
　Cytisus scoparins, 171–172
　Gompholobium huegelii, 152
　gorse, 172
　grapefruit, 15–16
　Lantana camara, 171
　Lunicera japonica, 172
　Lupinus arboreus, 171
　Lythrum salicaria, 173–174, 177
　macadamia, 16–17
　melon, 35, 85, 114t, 118, 185, 219
　native plants and effects of invasive weeds, 175, 178
　pear, 35
　pepper, 36–39
　squash, 174
　strawberry, 12, 28, 85
　star-thistle, yellow, 171, 174
　sunflower, 12, 15
　tomato, 13, 37, 55, 158–159, 219
　tree fruits, 51
　Ulex europeans. *See* pollination of, gorse
　watermelon, 11, 15, 19, 20, 36, 40, 85
　weeds, 166, 168–173, 177–178
　zucchini, 36, 40–41
pollinator biocontrol vector technology, 65–75
　using bees, 65–75
　dispenser designs, 72–74
　in greenhouses, 69–71
　for insect pests, 68–71
　for plant pathogens, 67–68
　safety, 71, 74–75
pollinator dependency factor, 28t
pollinators
　displacement of, 150, 152–153, 156, 159
　exotic, 157–159
　guilds, 52, 54–56, 58–61
　management, 60, 83
　needs of, 7–9
　services, 10–21, 151, 167–168, 171, 175, 205, 221
　status of, 167, 220
　stewardship of, 51, 55, 59, 61
polycarbonate, 30
polyethylene, 30
Polygonaceae, 50t
polylecty, 5
polyvinylchloride (PVC), 30, 109
population density, of plants, 167f, 168, 198
portfolio effect, 18
predator, 16, 20, 35, 59–60
prenesting period, 86f, 86–88, 94–95, 98

prepupae, (*see also* diapause)
　development of, 89–90, 109
　dormancy in, 85–86, 88–89, 92, 110–113, 135
prewintering period, 86f, 86, 91–93, 96–97
probiotics, 125
protease inhibitors, 208–209
protozoa, 127t, 128, 134
Pseudomasaris vespoides. See pollen wasp
Pseudomonas fluorescens, 68, 75

queen excluder, 134

raspberry, 67 (*see also* pollination of)
rehabilitation of rangeland/wildland, 49–50, 50t, 61
reproductive success
　of alfalfa leafcutting bees, 117
　of alkali bees, 115
　of bees, 8, 158
　of *Bombus occidentalis*, 149
　of *Cytisus scoparius*, 171–172
　of *Osmia*, 94
　of plants, 150, 167
respiration rate, 89–93, 97–98
restoration
　on/near farmland, 21, 213
　plant communities, 52, 178
　rangeland, 53f
　of sage grouse habitat, 50
　seed, 50–51, 61

safety
　of genetically modified plants, 184, 203–215
　of introduction of exotic bee species, 145–159
　of pesticides to bees, 16f, 115, 117, 215
　of pollinator biocontrol vector technology, 65–66, 71–72, 74–75
sagebrush, 48
sage grouse, 48, 50
Sapyga pumila, 116
second generation. *See* voltinism, multi

seed
　certification, 51, 106, 106t
　collection, 51
　mixes, 49, 61
　production, 41, 56, 75, 117, 177–178
　set, 12, 36, 38–39, 52, 56, 75, 98, 117, 153, 187–194, 211f, 211–212, 215
self-pollination. *See* pollination, self-
Selliera, 156
sex ratio, 5, 111, 114t
soil requirements, 109
Solanaceae, 28, 37–38, 69, 71, 158
source identified class, 51, 59
southeastern blueberry bee. *See* bee species, southeastern blueberry
Sphaeralcea
　ambigua, 50
　grossularifolia, 50t
　munroana, 50t
Spiroplasma sp. 126t
spore, 207
　fungal, 127–128, 133, 135–136, 136f
　vector, 65, 72–73
squash bee. *See* bee species, squash
star-thistle, yellow, 156 (*see also* pollination of)
stingless bees. *See* bee species, stingless
stone brood. *See* disease, of bees
strawberry, 28, 29t, 37t (*see also* pollination of)
　and biocontrol of pests, 67–68, 72, 74
Streptococcus pluton. See disease, of bees
sublethal effects
　of genetically modified plants, 214, 214f
　of pesticides on bees, 117, 206
sugar
　fructose, 28, 29t, 29, 38–39
　glucose, 28, 29t, 29, 38–39
　raffinose, 28, 29t
　stachyose, 28, 29t
　sucrose, 28, 29t, 38–40
sunflower, 17, 20, 59 (*see also* pollination of)
　and biocontrol of pests, 68–69
sweat bee. *See* bee species, sweat

synchrony
- asynchrony, 114, 114t
- of bloom and pollinator activity, 96, 98, 112

tarnished plant bug, 69–71, 70f
temperature
- and bee survival, 33
- and brood, 32, 71, 88–98, 109–110
- on crops, 106–107
- on diseases of bees, 127–128, 132f, 133
- and foraging behavior, 32, 36, 40
- in greenhouses, 32–33
- and management of stored bees, 84, 86–87, 114t
- and pollen and nectar production, 32, 39

threatened plants, 52
thrips, 70f, 71
tomato, 15, 28t, 28, 30–33, 37t, 37–38, 71, 158–159, 185, 219 (*see also* pollination of)
tracheal mite, 131, 219
Trialeurodes vaporariorum. *See* whitefly
Trichoderma
- biocontrol of gray mold, 68, 71, 75
- biocontrol of *Sclerotinia*, 68
- safety towards bees, 71

Trichodes ornatus, 116
Trifolium pratense, 153, 174

Ulocladium atrum, 68
ultraviolet
- absorbing plastics, 31f, 31
- blocking plastic films, 27, 30
- light, 30
- transmission, 30

Vaccinium ashei. *See* pollination of, blueberry
Varroa, 10
- biological control of, 72
- as a disease vector, 133
- impact on honey beekeeping, 10, 118, 124, 219
- pesticide resistance, 10, 131

Varroa destructor. *See* Varroa
vector technology. See pollinator biocontrol vector technology
viruses
- for biocontrol, 65, 68, 125
- as pathogens of bees, 115, 124, 126–127t, 133 (*see also* disease, of bees)

visitation (to flowers by bees)
- frequency, 29
- rate, 149, 152, 171, 176

voltinism
- multi, 97, 109, 111, 116
- parsi, 86
- semi, 86
- uni, 85–86, 86f

wasps as different from bees, 5
watermelon, 28t, 28, 29t, 37t, 39–40, 42 (*see also* pollination of)
weeds
- floral rewards and resources, 173–174, 176, 178
- impact on pollinators, 166–168, 173–175, 178, 210–213
- and nesting habitat, 174–175, 178
- and native plants, 175–178

western flower thrips. (*See* thrips)
whitefly, 71
winter period, 19, 57, 84–87, 89, 91–98, 109–111, 113, 116, 135
Wyethia, 60

Xylocopinae, 6

zucchini, 28t, 28, 29t, 30, 36, 37t (*see also* pollination of)

DATE DUE

GAYLORD PRINTED IN U.S.A.